Lecture Notes in Physics

Edited by H. Araki, Kyoto, J. Ehlers, München, K. Hepp, Zürich
R. Kippenhahn, München, H.A. Weidenmüller, Heidelberg,
J. Wess, Karlsruhe and J. Zittartz, Köln

Managing Editor: W. Beiglböck

312

Joel S. Feldman
Thomas R. Hurd
Lon Rosen
Jill D. Wright

"QED: A Proof of Renormalizability"

Springer-Verlag
Berlin Heidelberg GmbH

Authors

Joel S. Feldman
Thomas R. Hurd
Lon Rosen
Department of Mathematics, University of British Columbia
Vancouver, B.C., V6T 1Y4, Canada

Jill D. Wright
Department of Mathematics, University of Virginia
Charlottesville, VA 22903, USA

ISBN 978-3-662-13663-8 ISBN 978-3-540-45953-8 (eBook)
DOI 10.1007/978-3-540-45953-8

Originally published by Springer-Verlag Berlin Heidelberg New York in 1988
Softcover reprint of the hardcover 1st edition 1988

2158/3140-543210

Preface

The central mathematical problem of quantum field theory, as it is currently formulated in terms of Euclidean Feynman integrals, is to construct a measure on the space of fields $\{\Phi(x)\}$ given by

$$d\nu(\Phi) = \text{const.}e^{\lambda V(\Phi)}dP(\Phi) \ .$$

Here the Gaussian measure $dP(\Phi)$ is determined by the free or quadratic part $\mathcal{L}_0$ of the Lagrangian of the model; its covariance C is given by

$$\frac{1}{2}(\Phi,\ C^{-1}\Phi) = \int \mathcal{L}_0(\Phi(x))\ dx \ .$$

The potential $V(\Phi)$ is determined by the interaction part $\mathcal{L}_{int}$ of the Lagrangian:

$$V(\Phi) = -\int \mathcal{L}_{int}(\Phi(x))\ dx \ .$$

These measures may not be genuine for technical reasons - the presence of fermions or lack of regularity - and so what we ask of the above formal expression for $d\nu$ is that its moments exist.

The Gaussian measure $dP(\Phi)$ is well understood. For instance, the integral of a polynomial in Φ can be elegantly evaluated as a sum over graphs whose lines correspond to the covariance C. The non-Gaussian measure $d\nu(\Phi)$ is another matter. Postponing the question of the actual existence of $d\nu$, we can ask whether it exists in perturbation theory, i.e., as a formal power series in the coupling constant λ, again a question about Gaussian integrals. Although simpler, this question, or some version of it, has been under investigation for about half a century. The difficulties are well known: in all models of interest, the covariance $C(x,y)$ is a classical Green's function with short-distance or ultraviolet (UV) singularities as $|x-y| \to 0$; if massless fields are involved, there is in addition the long-distance or infrared (IR) problem that $C(x,y)$ does not decay exponentially as $|x-y| \to \infty$. As a result, most of the graphs in the perturbation series are infinite.

The folk remedy is to cancel these infinities by adjusting or renormalizing V with counterterms $\delta V(\Phi,\lambda)$:

$$d\nu \rightarrow d\nu_{ren} = \text{const.} e^{\lambda V + \delta V} dP .$$

The counterterms $\delta V(\Phi,\lambda)$ are permitted to have the same form as terms in the original Lagrangian, but the coefficients of these terms are formal power series in λ which themselves have infinite coefficients. The central problem of perturbative renormalization theory is to demonstrate that there is some choice of δV for which all the infinities cancel, yielding a renormalized perturbation series with finite coefficients.

Such a demonstration typically encounters severe combinatorial and graphical complexities. To each order in λ, there are many graphs. Elementary power counting considerations may indicate that a graph G is finite but such power counting is too superficial in that G may contain divergent subgraphs. So a good renormalization algorithm on a graph G must first make subtractions on the divergent subgraphs of G, beginning with the smallest. But - and this is the notorious problem of "overlapping divergences" - what if two divergent subgraphs $G_1, G_2 \subset G$ intersect and neither is a subgraph of the other? In general, the renormalization procedure on G_1 will disturb that on G_2, and vice versa. Furthermore, how can we be sure that the required subtractions on all of these graphs to all orders can be implemented by an a priori choice of δV?

It took many years and the heroic efforts of many people to chart a safe course through these difficulties. Some of the milestones in this journey were the original work on QED[1] by Feynman, Schwinger and Tomonaga, the refinements by Dyson, Matthews, Salam, ..., the Dyson-Weinberg Power Counting Theorem[2], the renormalization prescription of Bogoliubov and Parasiuk[3] and the subsequent improvements of Hepp[4] and Zimmermann[5], culminating in the 1960s in what is now known as BPHZ renormalization[6]. Still, in 1970, a student of perturbative renormalization knew that he was not embarking on a pleasure cruise.

The early 1970s brought a new set of ideas to the subject, namely the renormalization group ideas of Wilson[7]. As interpreted by Gallavotti and co-workers[8-10], this approach is based on making scale decompositions of the fields or of the covariance: $\Phi = \sum_{h=-\infty}^{\infty} \Phi^{(h)}$ or $C = \sum_{h=-\infty}^{\infty} C^{(h)}$, where $C^{(h)}$ has length scale M^{-h}, $M > 1$ being a fixed scale parameter. In effect, this decomposition resolves the UV and IR singularities of C. By successively integrating out the fields $\Phi^{(h)}$ (from high to low h), Gallavotti and Nicolò[10] obtained a natural and beautiful tree expansion for $d\nu_{ren}$. The GN tree expansion dramatically simplifies the problem of perturbative renormalization, enabling one to make a choice of counterterms δV and to renormalize scale by scale without ever seeing overlapping divergences or the usual combinatorial complexities. With the control of the GN tree expansion, it is then relatively easy to show that a renormalized graph is finite and to obtain a sharp estimate on its size.

If one wishes to apply the GN method to a gauge field model, a basic problem arises because the scale decompositions do not respect gauge invariance: the model is renormalizable but it is not clear that renormalization can be achieved using only gauge invariant counterterms δV. (For that matter, this problem arises in any renormalization scheme based on BPHZ ideas.) In this monograph we return to the original model of QED and verify that the GN method can be applied with only gauge invariant counterterms in $d\nu_{ren}$.

In this monograph we have tried hard to provide a complete exposition that will be accessible to a wide audience - not just to experts in field theory. In 1988, perturbative renormalization may still not be a pleasure cruise, but we believe that the student can look forward to a relatively easy journey which boasts a number of beautiful vistas.

Acknowledgement. It is a pleasure to thank Douglas Jameson for a superb typing job. In spite of countless revisions and the frustrations of labelling imaginary graphs on the screen, Doug could always be counted on to give us a beautiful typescript with his usual humour and a maximum of grumbling. Any errors in this monograph are his responsibility and not ours.

TABLE OF CONTENTS

§1. Introduction

In 1951 Matthews and Salam[11] formulated a requirement for renormalization procedures that has become popularly known as the "Salam Criterion":

> "The difficulty, as in all this work, is to find a notation which is both concise and intelligible to at least two persons, of whom one may be an author."

Possibly there are many proofs of the renormalizability of quantum electrodynamics (QED) which satisfy the Salam Criterion. But we must confess that none of us has yet qualified as that other person who is the guarantor of the Criterion. While there are today many standard texts[12,13] which discuss the renormalizability of QED, we are not aware of any which presents a complete proof and in particular justifies the claim that only gauge invariant counterterms are required. We here submit to you a direct and <u>complete</u> proof and we invite you to judge whether you can vouch for the Salam Criterion.

Our concise and intelligible notation is not ours but was invented by Gallavotti and Nicolò[9,10] for the purpose of renormalizing the ϕ^4 model. We hasten to assert that their beautiful ideas not only provide critical notation but represent a fundamental advance in the study of renormalization. (Similar ideas appear in the work of Magnen, Sénéor[14], Gawedzki, Kupiainen[15], de Calan and Rivasseau[16], and Polchinski[17].)

In this monograph we present a self-contained proof of the renormalizability of a general Euclidean quantum field theory based on the Gallavotti-Nicolò tree expansion. In Section 2 we consider the expansion in the ultraviolet regime and in Section 6 we extend the expansion to the infrared regime. The basic scheme for renormalization is as follows:

Consider a (Euclidean) field theory[18] in d dimensions with free fields $\Phi = (\Phi_1, \Phi_2, \ldots)$, free (gaussian) measure $dP(\Phi)$, and covariance

$$C_{ij}(x,y) = \int \Phi_i(x)\Phi_j(y)dP(\Phi) = \langle\Phi_i(x)\Phi_j(y)\rangle \ .$$

For example, in QED the basic fields are $\Phi_1 = A$, $\Phi_2 = \psi$, $\Phi_3 = \bar{\psi}$ where A is the photon field and ψ, $\bar{\psi}$ are the electron fields (we generally suppress indices). The covariance is

$$C(x,y) = \begin{pmatrix} \langle\Phi_1(x)\Phi_1(y)\rangle & 0 & 0 \\ 0 & 0 & \langle\Phi_2(x)\Phi_3(y)\rangle \\ 0 & \langle\Phi_3(x)\Phi_2(y)\rangle & 0 \end{pmatrix} = \begin{pmatrix} D(x,y) & 0 & 0 \\ 0 & 0 & S(x,y) \\ 0 & -S^T(x,y) & 0 \end{pmatrix}$$

where $D(x,y)$ is the photon propagator, $S(x,y)$ the electron propagator and $S^T(x,y)$ its transpose [see (1.11)-(1.12)]. Formally, the measure dP is given by

$$dP(\Phi) = \text{const.}e^{-1/2\Phi C^{-1}\Phi} \prod_i \mathcal{D}\Phi_i \tag{1.1}$$

where we take the appropriate precautions for the anticommuting fermi fields.

We decompose each field Φ_i into a sum of independent fields

$$\Phi_i = \sum_{h=-\infty}^{\infty} \Phi_i^{(h)} \tag{1.2}$$

where the $\Phi^{(h)}$'s are free fields whose length scale is M^{-h}, $M > 1$ being a fixed scale parameter. More precisely, that the different $\Phi^{(h)}$'s are independent means that $\int \Phi_i^{(h)}\Phi_j^{(k)} dP(\Phi) = 0$ for $h \neq k$; that $\Phi_i^{(h)}$ has length scale M^{-h} means that the propagator

$$C_{ij}^{(h)}(x,y) = \int \Phi_i^{(h)}(x)\Phi_j^{(h)}(y)dP(\Phi)$$

satisfies [see (1.13)-(1.15)]

$$|C_{ij}^{(h)}(x,y)| \leq \text{const.}M^{2\delta_i h} e^{-M^h|x-y|} \tag{1.3}$$

where δ_i is the dimension of the field Φ_i ($\frac{d-2}{2}$ for bose and $\frac{d-1}{2}$ for fermi fields). The effect of the expansion (1.2) is to resolve the ultraviolet (UV) singularity of $C(x,y) = \sum_h C^{(h)}(x,y)$ as $x-y \to 0$ and the infrared (IR) singularity as $x-y \to \infty$ into a sum over scales.

UV-regularized fields are defined by truncating the upper range of scales in (1.1): $\Phi^{(\leq U)} = \sum_{h\leq U} \Phi^{(h)}$; the IR-regularized fields are $\Phi^{(\geq I)} = \sum_{h\geq I} \Phi^{(h)}$. We also write $\Phi^{[I,U]} = \sum_{h=I}^{U} \Phi^{(h)}$ for the fully regularized field, and $\Phi^{(k,U]} = \sum_{h=k+1}^{U} \Phi^{(h)}$, etc. Massive fields are already regular in the infrared: We assume for convenience that the bare masses $m_i \geq 1$, and then the decomposition (1.1) begins at h=0 (or $\Phi_i^{(h)} \equiv 0$ for $h < 0$).

Suppose that we are given a model with interaction $-V(\Phi)$, a local functional of $\Phi(x)$. Formally, the generating functional of connected Green's functions (with external lines amputated by C^{-1}) is given by

$$G(\Phi^e) = \log \frac{1}{Z} \int e^{\Phi^e C^{-1} \Phi} e^{V(\Phi)} dP(\Phi), \tag{1.4}$$

where $Z = \int e^{V(\Phi)} dP(\Phi)$ and the source field or external field $\Phi^e = (\Phi_1^e, \Phi_2^e, \ldots)$ is suitably regular and has the same nature as Φ (in particular, Φ_i^e is fermionic if Φ_i is). Removing the free field contribution and changing variables ($\Phi \rightarrow \Phi + \Phi^e$) we find from (1.1) that

$$V_{e,un}(\Phi^e) \equiv G(\Phi^e) - \frac{1}{2} \Phi^e C^{-1} \Phi^e = \log \frac{1}{Z} \int e^{V(\Phi+\Phi^e)} dP(\Phi).$$

$V_{e,un}$ is the (unrenormalized) <u>effective potential</u> for the model. To make sense of $V_{e,un}$ we regularize and renormalize V with counterterms $\delta V^{I,U}$ of the same form as the terms in the Lagrangian (free or interaction):

$$V_e^{I,U}(\Phi^e) \equiv \log \int e^{(V+\delta V^{I,U})(\Phi^{[I,U]}+\Phi^e)} dP(\Phi^{[I,U]}) + \text{const.} \tag{1.5}$$

where the constant is chosen to cancel the terms independent of Φ^e arising from the integration. (Technically, this constant is infinite because of the volume divergence, but this problem is easily cured by imposing and then removing a volume cutoff, and so we shall continue to indicate the removal of these constant terms by writing "+ const."). Note that the effective potential (1.5) is distinct from the more usual definition which is given by a Legendre transform[19] and generates the one-particle-irreducible amputated Green's functions.

If it is possible to choose allowed counterterms such that

$$V_e = \lim_{\substack{U\to\infty \\ I\to-\infty}} V_e^{I,U} \tag{1.6}$$

exists then we say that the model is <u>renormalizable</u>. In the context of perturbation theory, this means that the limit (1.6) exists for each order in the expansion of $V_e^{I,U}$ in powers of the coupling constant(s).

We consider also the effective potential at scale k obtained by integrating out as in (1.5) all the fields with scale > k:

$$V_k^{I,U}(\Phi^{(\leq k)}) \equiv \log \int e^{(V+\delta V^{I,U})(\Phi^{[I,U]})} dP(\Phi^{(k,U]}) + \text{const.} \tag{1.7}$$

For models with no IR divergences or with a fixed IR cutoff I the existence of the limit (for each fixed k)

$$V_k^I = \lim_{U\to\infty} V_k^{I,U}$$

amounts to UV-renormalizability; in general, the existence of the limit

$$\lim_{k\to-\infty} \lim_{I\to-\infty} V_k^I = V_e \tag{1.8}$$

amounts to IR- and UV-renormalizability.

The difficulty, as in all this work, is to choose the counterterms $\delta V^{I,U}$ in such a way as to exhibit the renormalization cancellations in $V_k^{I,U}$, uniformly in I,U. The key to the Gallavotti-Nicolò approach is a representation for $V_k^{I,U}$ as a sum (2.18) over "connectivity trees" τ:

$$V_k^{I,U} = \sum_\tau V_{k,\tau}^{I,U} . \tag{1.9}$$

This sum arises naturally if in (1.7) we integrate out the fields $\Phi^{(U)}, \Phi^{(U-1)}, \ldots, \Phi^{(k+1)}$ in succession. For instance integrating out $\Phi^{(U)}$ represents $V_{U-1}^{I,U}$ as a sum of graphs with "external fields" $\Phi^{(\leq U-1)}$ and internal connecting lines $C^{(U)}$. Integrating out $\Phi^{(U-1)}$ produces graphs with external fields $\Phi^{(\leq U-2)}$ and additional connecting lines $C^{(U-1)}$, and so on. The sum (1.9) organizes the graphs produced by integrating out $\Phi^{(k,U]}$, according to the scales at which the connections are made. Each $V_{k,\tau}^{I,U}$ itself is a sum of different connected Feynman graphs whose connections are "consistent" with τ. The graphs which require renormalization are those whose net dimension in the fields $\Phi^{(\leq k)}$ is d or less. Counterterms are introduced scale by scale, in descending order, as required to renormalize the effective potential at that scale. The resulting straightforward estimates amount to standard power counting.

The advantages of the GN procedure are:

1. The notorious complication of "overlapping divergences"[20] never enters, since the tree expansion automatically decomposes graphs into the appropriate "Hepp sectors".
2. No appeal to separate power counting theorems is required.
3. The renormalization cancellations are made directly in the exponent without first expanding. This results in an algebraic simplification and clarifies the connection between renormalization and counterterms in the Lagrangian.

We now describe how we adapt the GN procedure to QED_4. By QED we mean Euclidean quantum electrodynamics in 4 dimensions. In Appendix B we describe the relation between our renormalization of the Euclidean model and the renormalization of the objects of physical interest (Green's functions and scattering amplitudes) in the relativistic model. In terms of the Euclidean fields $\Phi_1 = A_\mu$, $\Phi_2 = \psi$, $\Phi_3 = \bar{\psi}$, the Lagrangian density is given by

$$\begin{aligned}\mathcal{L} &= \mathcal{L}_p + \mathcal{L}_f + \mathcal{L}_{int.} \\ &= \left[\frac{1}{4} F^2 + \frac{\lambda}{2} (\partial . A)^2\right] + \bar{\psi}(-i\not{\partial} + m)\psi + e\bar{\psi}\not{A}\psi \,. \end{aligned} \tag{1.10}$$

Here (repeated indices are summed over)

$$F^2 = F_{\mu\nu}F_{\mu\nu}, \qquad F_{\mu\nu} = \partial_\mu A_\nu - \partial_\nu A_\mu$$

$$\partial . A = \partial_\mu A_\mu$$

$$\not{A} = A_\mu \gamma^\mu, \qquad \not{\partial} = \gamma^\mu \partial_\mu$$

in terms of the 4×4 antihermitian Euclidean Dirac matrices, which satisfy

$$\gamma^\mu\gamma^\nu + \gamma^\nu\gamma^\mu = -2\delta^{\mu\nu},$$

$m > 0$ is the bare electron mass, e the bare electron charge and $\lambda > 0$ the gauge fixing parameter.

In momentum space the quadratic form

$$\int \mathcal{L}_p dx = \frac{1}{2} \int \overline{A_\mu(k)} \left[k^2\delta_{\mu\nu} + (\lambda-1)k_\mu k_\nu\right] A_\nu(k) dk.$$

The simplest choice of gauge parameter is $\lambda = 1$ (Feynman gauge) and we accordingly fix $\lambda = 1$. The resulting free photon measure

$$d\mu(A) = \text{const.}\, e^{-\int \mathcal{L}_p dx} \mathcal{D}A$$

has covariance or propagator

$$D_{\mu\nu}(x,y) = \int A_\mu(x) A_\nu(y) d\mu(A)$$

$$= \delta_{\mu\nu} \frac{1}{(2\pi)^4} \int \frac{e^{ik(x-y)}}{k^2} dk. \tag{1.11a}$$

The free fermi measure is

$$d\nu(\psi,\bar{\psi}) = \text{const.}\, e^{-\int \mathcal{L}_f dx} \mathcal{D}\psi \mathcal{D}\bar{\psi}$$

with propagator (we suppress spinor indices)

$$S(x,y) = \int \psi(x)\bar{\psi}(y) d\nu$$

$$= \frac{1}{(2\pi)^4} \int (\not{k}+m)^{-1} e^{ik(x-y)} dk. \tag{1.11b}$$

Notwithstanding the fact that $d\nu$ involves anticommuting variables and so is not a true gaussian measure (see e.g. Berezin[21]), we have in terms of the above general formalism:

$$dP(\Phi) = d\mu(A) d\nu(\psi,\bar{\psi})$$

$$V(\Phi) = -\int \mathcal{L}_{int} dx = -e \int :\bar{\psi}(x) \not{A}(x) \psi(x): dx. \tag{1.12}$$

The expansion of the electron propagator over scales is

$$S(x,y) = \sum_{h=-\infty}^{\infty} S^{(h)}(x,y) \tag{1.13a}$$

where $S^{(h)} = 0$ if $h < 0$,

$$S^{(h)}(x,y) = \frac{1}{(2\pi)^4} \int dk (\not{k}+m)^{-1} e^{ik(x-y)} \left[e^{-(k^2+m^2)/M^{2h}} - e^{-(k^2+m^2)/M^{2h-2}} \right] \tag{1.13b}$$

if $h>0$ and

$$S^{(0)}(x,y) = \frac{1}{(2\pi)^4} \int dk (\not{k}+m)^{-1} e^{ik(x-y)} e^{-(k^2+m^2)}. \tag{1.13c}$$

That of the photon propagator is

$$D(x,y) = \sum_{h=-\infty}^{\infty} D^{(h)}(x,y) \tag{1.14a}$$

where

$$D^{(h)}(x,y) = \frac{1}{(2\pi)^4} \int dk\, k^{-2} e^{ik(x-y)} \left[e^{-k^2/M^{2h}} - e^{-k^2/M^{2h-2}} \right]. \tag{1.14b}$$

With these decompositions it is easy to verify the bounds (1.3) (see (3.22)-(3.25)):

$$|\partial_x^n S^{(h)}(x,y)| \leq \text{const. } M^{(3+|n|)h} e^{-M^h|x-y|} \tag{1.15a}$$

and

$$|\partial_x^n D^{(h)}(x,y)| \leq \text{const. } M^{(2+|n|)h} e^{-M^h|x-y|} \tag{1.15b}$$

where $\partial_x^n = \prod_{i=1}^{4} \left(\frac{\partial}{\partial x_i}\right)^{n_i}$ and $|n| = \sum_{i=1}^{4} n_i$. (For notational convenience we have scaled the unit of length so that $m \geq 1$ and (1.15a) holds for $h = 0$.) The electron propagator with UV cutoff $U > 0$ is $S^{(\leq U)} = \sum_{h=0}^{U} S^{(h)}$, and the photon propagator with UV cutoff $U > 0$ and IR cutoff $I \leq 0$ is $D^{[I,U]} = \sum_{h=I}^{U} D^{(h)}$.

The renormalization of QED would be a routine application of the GN expansion were it not for the complication of gauge invariance. The local, Euclidean invariant terms of dimension 4 or less that arise as counterterms are

$$F^2,\ \bar{\psi}\psi,\ -i\bar{\psi}\partial\!\!\!/\psi,\ \bar{\psi}A\!\!\!/\psi \tag{1.16}$$

and

$$(\partial\cdot A)^2,\ A^2,\ A^4. \tag{1.17}$$

However, the terms in (1.17) are gauge-variant and are forbidden in the renormalization of QED. Inasmuch as the GN procedure is inherently gauge variant we cannot rule out these forbidden counterterms.

To understand this dilemma let us examine the Ward[22] identities. In terms of the (formal) expectation (1.4) and (1.12), define

$$W(\eta,\bar{\eta},B) = \log \frac{1}{Z} \int e^{-\int dx(e\bar{\psi}\not{B}\psi + \bar{\psi}\eta + \bar{\eta}\psi)} e^{V(\Phi)} dP(\Phi)$$

where B_μ is a vector source and η and $\bar{\eta}$ spinor sources. Since $S^{-1} = -i\not{\partial}+m$ satisfies

$$e^{-ie\chi} S^{-1} e^{ie\chi} = S^{-1} + e\not{\partial}\chi \tag{1.18a}$$

we have

$$d\nu(e^{ie\chi}\psi,\ e^{-ie\chi}\bar{\psi}) = e^{-e\int \bar{\psi}\not{\partial}\chi\psi} d\nu(\psi,\bar{\psi}) \tag{1.18b}$$

and so

$$W(\eta,\bar{\eta},B) = W(e^{-ie\chi}\eta,\ \ e^{ie\chi}\bar{\eta},\ \ B + \partial\chi)\ . \tag{1.19}$$

These are the Ward identities. Note that the Ward identities depend on the relations (1.18) for $d\nu(\psi,\bar{\psi})$ and not on properties of $d\mu(A)$ (which is in any case not gauge invariant since $\mathcal{L}_p$ has a gauge fixing term).

Suppose we knew that the counterterms $\delta V^{I,U}(\psi,\bar{\psi},A)$ satisfied the Ward identities (1.19):

$$\delta V^{I,U}(\psi,\bar{\psi},A) = \delta V^{I,U}(e^{-ie\chi}\psi,\ \ e^{ie\chi}\bar{\psi},\ \ A + \partial\chi).$$

Then we could conclude that the forbidden counterterms (1.17) do not occur and that only the counterterms (1.16) occur (with the last two in the combination $-i\bar{\psi}\not{\partial}\psi + e\bar{\psi}\not{A}\psi$). On the other hand if we regularize the fermi propagator S then S^{-1} is no longer a simple first order differential operator satisfying (1.18a), and we lose the Ward identities as well as the gauge invariance of $\mathcal{L}$. But the GN expansion is based upon regularizing and making a scale decomposition of every propagator, in particular, of S:

$$S^{(\leq N)} = \sum_{h=0}^{N} S^{(h)}. \tag{1.20}$$

Thus there is a basic conflict between the GN procedure and the requirements of gauge invariance.

To resolve this conflict we introduce an auxiliary regularization involving S which does not disturb the Ward identities. The regularization we choose is "loop regularization"[23], implemented by means of fictitious spinor fields ψ_i, $\bar{\psi}_i$, $i = 1,2,3$. For the (standard) details, see Section 3. In this regularization, ψ_1 is a fermi field, ψ_2 and ψ_3 are bose fields, and the Lagrangian $\mathcal{L}$ in (1.10) is replaced by

$$\mathcal{L}_\Lambda = \mathcal{L} + \sum_{i=1}^{3} \bar{\psi}_i(-i\not{\partial} + M_i + e\not{A})\psi_i \tag{1.21}$$

where $M_1^2 = m^2 + 2\Lambda^2$, $M_2^2 = M_3^2 = m^2 + \Lambda^2$ so that

$$M_1^2 + m^2 = M_2^2 + M_3^2. \tag{1.22}$$

The effect of these extra terms is to replace each fermi loop occurring in the effective potential by a second difference of loops. When $\Lambda < \infty$ and when the photon propagator is also regularized ($U < \infty$ and $I > -\infty$) the computation of the (unrenormalized) effective potential then yields finite results in perturbation theory, i.e. the contribution of each order of perturbation theory is finite (Lemma 3.1).

Unfortunately, the loop regularization cannot be used to define a sufficiently sensitive scale decomposition of the fermi propagator. The difficulty is that one cannot decompose each fermi propagator in a loop independently, as is necessary for the introduction of the correct counterterms. Accordingly, we still decompose S as in (1.20) (with $N = \infty$) and we are obliged to accept counterterms at each scale that are not gauge invariant. However, when the gauge variant counterterms are summed up over all scales they add up to zero by virtue of the Ward identities (see Section 4 for this argument). In this way we can define a renormalized effective potential $V_k^{I,U,\Lambda}$ in which only gauge invariant counterterms are employed but in which cancellations are performed in a gauge variant way. This strategy, which we believe is essential in any renormalization scheme for QED based on power counting and BPHZ ideas, seems not to have been followed in the standard texts.

We remark that in place of the auxiliary loop regularization we could have deployed any other regularization which preserves the Ward identities, such as dimensional regularization[24]. We have chosen loop regularization because of its simplicity in the functional integral formalism for the continuum theory. However, loop regularization brings with it a technical problem: In renormalizing $V_k^{I,U,\Lambda}$ we use the correct counterterms ($\bar{\psi}\psi$, $\bar{\psi}\partial\!\!\!/\psi$, $\bar{\psi}A\!\!\!/\psi$) for the original spinor fields but not for the fictitious fields ψ_i (see (3.31)). The reason is that the loop regularization is a rather delicate subtraction that works only if the masses satisfy (1.22) and if the terms $\bar{\psi}(-i\partial\!\!\!/ + eA\!\!\!/)\psi$ and $\bar{\psi}_i(-i\partial\!\!\!/ + eA\!\!\!/)\psi_i$ in (1.21) occur with the same coefficient. The correct counterterms for ψ_i would spoil this subtraction.

As a consequence it is essential that the Λ regularization be removed before I and U. With $I > -\infty$ and $U < \infty$ fixed, the graphs in $V_k^{I,U,\Lambda}$ that appear to require fictitious field counterterms $\bar{\psi}_i\psi_i$, $\bar{\psi}_i\partial\!\!\!/\psi_i$ and $\bar{\psi}_iA\!\!\!/\psi_i$ are actually finite. In other words, these counterterms are not needed for renormalization cancellations but only for loop subtractions. Hence, as $\Lambda \to \infty$ and $M_i(\Lambda) \to \infty$, we obtain the convergence $V_k^{I,U,\Lambda} \to V_k^{I,U}$, where $V_k^{I,U}$ contains no reference to fictitious fields (see Section 5).

Our basic strategy may be summarized as follows. We first introduce four cutoffs: N (an UV cutoff on the electron propagator), Λ (an UV cutoff on electron loops), U (an UV cutoff on the photon propagator), and I (an IR cutoff on the photon propagator). With all four cutoffs in place we choose cutoff dependent counterterms (including gauge variant counterterms) and, using the GN formalism, we establish bounds on each order of the resulting renormalized perturbation expansion for the effective potential. We then take $N \to \infty$ (Theorem 3.2) and prove that in this limit the gauge variant counterterms sum to zero, order by order in perturbation theory (Theorem 4.1). We then take $\Lambda \to \infty$, proving that in this limit the fictitious fields disappear and that the "wrong" choice of fictitious field counterterms is harmless (Theorem 5.1). Next (Theorem 5.3) comes the limit $U \to \infty$ (UV-renormalizability). The proof that $V_k^{I,U}$ converges to V_k^{I} amounts to standard power counting. Finally, in Section 7 we remove the IR cutoff and establish the renormalizability (1.8) of QED.

The results of Sections 2-7 are all statements in perturbation theory, i.e. statements about the coefficients $V_{e,n}$ in the formal power series $\sum_{n=1}^{\infty} V_{e,n} e^n$ for the effective potential V_e. However as we show in Section 8 for a general model, our methods lead to large order bounds[16] on the coefficients which imply the local Borel summability of the formal power series (when the degree of the interaction is 4 or less). For QED these bounds are of the form

$$|V_{e,2n}| \leq \text{const. } n!\ R^{-n}$$

where $R > 0$. Hence the Borel-transformed series $\sum_{n=1}^{\infty} V_{e,2n} \alpha^n / n!$, in terms of the fine structure constant $\alpha = e^2/4\pi$, has a positive radius of convergence R.

The analysis of Sections 2-8 takes place in the Euclidean world. Once the Euclidean Green's functions of a model have been constructed and shown to satisfy appropriate axioms, then an Osterwalder-Schrader type reconstruction theorem[34] can be invoked to return to the relativistic world and real time. However, inasmuch as our results are statements in perturbation theory, the question remains: what, if anything, do our renormalization procedure and our bounds have to do with the real time perturbation series? Appendix B ("Real Time") is devoted to this question. We show there that the same UV-renormalization scheme works in the relativistic world, albeit with a surprising increase in technical difficulty because of the lack of exponential decay of the propagators in relativistic x-space. Our analysis takes place in "α-p-space" but relies on bounds back in Euclidean x-space.

In summary, in this monograph we give a complete account of the Gallavotti-Nicolò renormalizability method for a general model in both the UV and IR regimes, and we provide an unambiguous proof of the renormalizability of QED. In doing so we have shown that the Gallavotti-Nicolò formalism is applicable to gauge field theory.

§2. The GN Tree Expansion and UV-Renormalization

We present in this section a version of the GN method of UV-renormalization as it applies to a general strictly renormalizable model. While all the essential ideas can be found in the papers of Gallavotti and Nicolò[9,10], our treatment is a substantial reworking of their method. For a model like QED with IR divergences, we implicitly assume throughout this section (and until §6) that we have imposed an IR cutoff I = 0. In §6 we extend the GN method to the IR regime.

If we allow gauge variant counterterms, the renormalization of QED is a straightforward application of the methods described in §§2 and 6. The modifications required to establish that only gauge invariant counterterms are needed in QED will be given in §§3, 4, 5 and 7.

For your convenience we have compiled a glossary of the numerous definitions and notations of the GN expansion in Appendix A.

The unrenormalized expansion

Consider a Euclidean field theory whose fields $\Phi(x) = (\Phi_1(x), \Phi_2(x), \ldots)$ have been decomposed into independent fields $\Phi^{(h)}$ of different length scales, as described in (1.2)-(1.3) and whose interaction is given by $-V(\Phi)$.

We impose an UV cutoff on the theory by replacing Φ by $\Phi^{(\leq U)}$ for some fixed integer U>0. For the cutoff model, the unrenormalized potential is given by

$$V^U_{U,un}(\Phi^{(\leq U)}) \equiv V(\Phi^{(\leq U)}) \tag{2.1}$$

and the unrenormalized effective potential at scale $k \geq 0$, $V^U_{k,un}$, may be defined inductively by integrating out scale components $\Phi^{(U)}$, $\Phi^{(U-1)}$, ..., $\Phi^{(k+1)}$ of $\Phi^{(\leq U)}$. Thus for $0 \leq k < U$

$$V^U_k = \log \int e^{V^U_{k+1}(\Phi^{(\leq k+1)})} dP(\Phi^{(k+1)}) + \text{const.} \tag{2.2}$$

where the constant term is chosen to cancel the terms independent of $\Phi^{(\leq k)}$ in the first term. For notational simplicity we shall omit the subscript "un", although until (2.37) we shall be dealing exclusively with the unrenormalized effective potential.

We shall show that V_k^U is given by a sum of Feynman graphs with external legs corresponding to the unintegrated fields $\Phi^{(\le k)}$, with internal lines corresponding to propagators $C^{(h)}$, but with restrictions on the scales of these propagators. The GN expansion organizes these restricted graphs in a systematic way that permits efficient estimates.

In the GN method, it is always the exponent V_k^U of the path integral for the effective theory at a given scale that is evaluated. We review this calculation in detail for the case where a single scale is integrated out. Repeated application of the same ideas will yield the general case.

V_{U-1}^U is found from $V_U^U = V$ as follows:

$$V_{U-1}^U\;(\Phi^{(\le U-1)}) = \log \mathcal{E}_U(e^{V(\Phi^{(\le U)})}) + \text{const.} \tag{2.3}$$

where for any integer h,

$$\mathcal{E}_h(.) = \int .\; dP(\Phi^{(h)}). \tag{2.4}$$

Hence by Taylor's Theorem,

$$\begin{aligned} V_{U-1}^U(\Phi^{(\le U-1)}) &= \sum_{n=1}^{\infty} \frac{1}{n!}\, \partial_\lambda^n \log \mathcal{E}_U(e^{\lambda V})\,|_{\lambda=0} \\ &= \sum_{n=1}^{\infty} \frac{1}{n!} \mathcal{E}_U^T(V,\ldots,V) \qquad \text{(n arguments)} \end{aligned} \tag{2.5}$$

defining $\mathcal{E}_U^T$, the <u>truncated expectation</u>. More generally, for n distinct arguments $X_1(\Phi^{(\le h)}),\ldots,X_n(\Phi^{(\le h)})$ we define

$$\mathcal{E}_h^T(X_1,\ldots,X_n) = \frac{\partial^n}{\partial\lambda_1\ldots\partial\lambda_n} \log \mathcal{E}_h(e^{\lambda_1 X_1+\ldots+\lambda_n X_n})\,|_{\lambda_j=0}.$$

Note that each X_i is a function of the fields $\Phi^{(\le h)}$, and the result is a function of $\Phi^{(\le h-1)}$.

The truncated expectation is evaluated using scale-restricted Feynman graphs.

Consider the example

$$\mathcal{E}_U^T(:\Phi^{(\leq U)}(x_1)^{p_1}:,\ldots,:\Phi^{(\leq U)}(x_n)^{p_n}:). \tag{2.6}$$

(For a discussion of Wick powers, see for example Ref. 9.)

We are suppressing indices but bear in mind that a power Φ^p stands for $\prod_i \Phi_i^{p_i}$ with a specific order of factors if anticommuting fields are involved. Now each $\Phi^{(\leq U)}$ in (2.6) can be written as $\Phi^{(U)} + \Phi^{(<U)}$ (designated hard + soft) and correspondingly $\mathcal{E}_U^T$ may be written as a sum of $2^{\Sigma|p_j|}$ terms. Since the soft components are unaffected by $\mathcal{E}_U^T$ and are independent of the hard components of the field, they may be extracted to form a product of Wick monomials multiplying each term. Thus each term is of the form

$$\prod_{j=1}^{n} :\Phi^{(<U)}(x_j)^{k_j}: \mathcal{E}_U^T(:\Phi^{(U)}(x_1)^{\ell_1}:,\ldots,:\Phi^{(U)}(x_n)^{\ell_n}:). \tag{2.7}$$

We now give a brief explanation of how (2.7) is evaluated in terms of graphs, brief because it is all fairly standard.

An ordinary expectation may be evaluated as a sum over graphs by the usual rules of "gaussian" integration:

$$\mathcal{E}_h\left(\prod_{j=1}^{n} :\Phi^{(h)}(x_j)^{\ell_j}:\right) = \sum_G V(G). \tag{2.8a}$$

Each graph G has vertices $x_1,\ldots,x_n$; from x_j there emerge $|\ell_j|$ legs or half-lines each corresponding to a field $\Phi_i^{(h)}$ in $\Phi^{(h)}(x_j)^{\ell_j}$; any two legs from different vertices may be paired to form a line of G. The sum over G in (2.8) is the sum over all such possible pairings that use up all the legs. The value V(G) of G is the product over the lines $\mathcal{L}(G)$ of G of the propagators associated with each line:

$$V(G) = \pm \prod_{\ell\in\mathcal{L}(G)} C_\ell^{(h)},$$

where $C_\ell^{(h)} = C_{i_\ell,j_\ell}^{(h)}(x_\ell,y_\ell)$, x_ℓ,y_ℓ are the two vertices at the ends of ℓ, and i_ℓ,j_ℓ are the types of the legs which form ℓ. The sign is ± depending on whether an

$\left\{\begin{matrix}\text{even}\\ \text{odd}\end{matrix}\right.$ number of commutations of fermi fields are required to make the pairings.

A truncated expectation, such as that occurring in (2.7), is evaluated in the same way as (2.8a) except that only <u>connected</u> graphs G occur in the sum (see, for example, Ref. 8):

$$\mathcal{E}_h^T(:\Phi^{(h)}(x_1)^{\ell_1}:, \ldots, :\Phi^{(h)}(x_n)^{\ell_n}:) = \sum_{G \text{ connected}} V(G) \tag{2.8b}$$

where the value V(G) has the same meaning as in (2.8a).

The product of soft Wick monomials in (2.7) can be rewritten as a sum of Wick monomials:

$$\prod_{j=1}^{n} :\Phi^{(<U)}(x_j)^{k_j}: = \sum_G V(G).$$

The sum takes place over all graphs G with vertices $x_1,\ldots,x_n$, $|k_j|$ legs emerging from each x_j, and lines formed by pairing any two legs from different vertices. However, not all legs need be used up to form lines, the unpaired legs being referred to as <u>external</u> legs of G. The value is

$$V(G) = \pm \prod_{\ell\in\mathcal{L}(G)} C_\ell^{(<U)} : \prod_{\lambda\in\Lambda(G)} \Phi_{i_\lambda}^{(<U)}(x_\lambda):$$

where $\Lambda(G)$ is the set of external legs of G, x_λ is the vertex from which λ emerges and i_λ is the type of λ.

Putting together the soft and hard factors contributing to each term of the form (2.7), we find that the truncated expectation (2.6) is given by a sum over graphs G on n vertices $x_1\ldots,x_n$, satisfying the following conditions:

(1) for each j, $|p_j|$ <u>legs</u> emerge from x_j;

(2) any two legs from different x_j's may be paired to form a <u>line</u> $\ell\in\mathcal{L}(G)$, and the unpaired legs are called the <u>external legs</u> $\Lambda(G)$ of G;

(3) all lines are designated hard or soft in such a way that the graph is connected by hard lines.

The <u>value</u> V(G) of each graph is the product of the following factors:

i) a Wick monomial in the external fields $:\prod_{\lambda\in\Lambda(G)} \Phi_{i_\lambda}^{(\leq U-1)}(x_\lambda):$,

ii) a factor $C_{i_\ell,j_\ell}^{(U)}(x_\ell,y_\ell)$ $\left(C_{i_\ell,j_\ell}^{(<U)}(x_\ell,y_\ell)\right)$ for each hard (soft) line $\ell\in\mathcal{L}(G)$,

iii) a ± sign depending on whether the pairing of fermion legs involved an even or odd number of commutations.

If we choose to identify topologically similar graphs, as we do, there is in addition:

iv) a combinatorial factor arising from the number of ways of designating each field $\Phi^{(\leq U)}$ as hard or soft, and the number of ways of joining legs from different vertices.

For example, with one species of field, $n=2$, $p_1=p_2=4$, 2 external legs and two soft lines, the graph

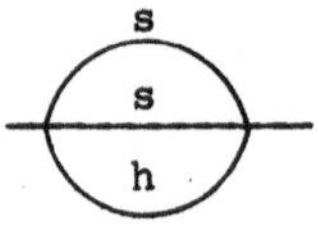

occurs in $\mathcal{E}^T(\times,\times)$ with a combinatorial factor $4^2 3!3$ arising from the number of ways of selecting the external fields (4^2), the number of ways of pairing the legs to form lines $(3!)$, and the number of ways of designating one line as hard (3).

Returning to the truncated expectation in (2.5), suppose that V is a sum of Wick monomials

$$V = \sum_\alpha V_\alpha = \sum_\alpha \lambda_\alpha \int :\Phi^{(\leq U)}(x)^{p_\alpha}:dx.$$

Then

$$\begin{aligned}\mathcal{E}^T_U(V,..,V) &= \sum_{\alpha_1,\dots,\alpha_n} \lambda_{\alpha_1}\dots\lambda_{\alpha_n} \int\dots\int dx_1\dots dx_n\ \mathcal{E}^T_U(:\Phi^{(\leq U)}(x_1)^{p_{\alpha_1}}:,\dots,:\Phi^{(\leq U)}(x_n)^{p_{\alpha_n}}:)\\ &= \sum_G V(G)\end{aligned}$$

where now the sum over graphs involves a sum over <u>types</u> α of vertices, and the value V(G) includes a "coupling constant" λ_α for each vertex of type α and an integration over the positions of the vertices. It is no problem to accommodate differentiated fields in the interaction V: $\partial^n_x \Phi_i$ simply enters the power counting with dimension $\delta_i + |n|$. For simplicity, we assume that no differentiated fields are present.

The evaluation of the effective potential V^U_k may now be described inductively. Suppose that V^U_{k+1} has been evaluated as a sum of (integrals of) Wick monomials

$$V^U_{k+1}\left(\phi^{(\le k+1)}\right) = \sum_i X_i\left(\phi^{(\le k+1)}\right).$$

Then by (2.2) and (2.5)

$$V^U_k\left(\phi^{(\le k)}\right) = \sum_{n=1}^{\infty} \frac{1}{n!} \mathcal{E}^T_{k+1}(V^U_{k+1},\dots,V^U_{k+1}) \qquad \text{(n arguments)}$$

$$= \sum_{n=1}^{\infty} \frac{1}{n!} \sum_{i_1,\dots,i_n} \mathcal{E}^T_{k+1}(X_{i_1},\dots,X_{i_n}). \tag{2.9}$$

We compute the truncated expectations $\mathcal{E}^T_{k+1}$ in terms of graphs just as we computed $\mathcal{E}^T_U$; i.e., vertices correspond to the X_i's, external legs to external fields $\phi^{(\le k)}$, hard lines to $C^{(k+1)}$, and soft lines to $C^{(<k+1)}$. The "generalized vertices" X_i are connected by the hard lines $C^{(k+1)}$.

In this way we find that V^U_r is expressed as a sum of nested truncated expectations, U-r layers deep. We just need to keep track of all of the combinatoric possibilities. The GN "tree" notation achieves this. In a GN tree, a truncated expectation $\mathcal{E}^T_h\left(X_1(\phi^{(\le h)}),\dots,X_p(\phi^{(\le h)})\right)$ with $p > 1$, is represented by a <u>fork</u> f with scale label h, and with p branches corresponding to the p arguments:

$$\mathcal{E}^T_h(X_1,\dots,X_p) \quad \leftrightarrow$$

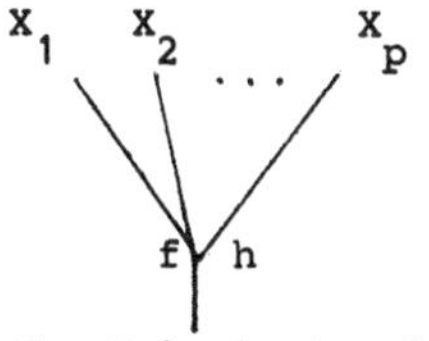

(2.10)

As we see from (2.9), each of the X_i's is itself a truncated expectation at scale h+1, and the fork (2.10) feeds into a truncated expectation at scale h-1. In this way, a tree of forks is built up. When p=1, the truncated expectation is just equal to an ordinary expectation. If we have a succession of ordinary expectations following a truncated expectation,

$$\mathcal{E}_{(k,h)}\left(\mathcal{E}^T_h(X_1,\dots,X_p)\right) \equiv \mathcal{E}_{k+1}\mathcal{E}_{k+2}\cdots\mathcal{E}_{h-1}\mathcal{E}^T_h(X_1,\dots,X_p)$$

we just picture it as a <u>segment</u> from k up to h

$$\xi_{(k,h)}\xi_h^T(X_1,\ldots,X_p) \quad \leftrightarrow \quad (k < h) \qquad (2.11)$$

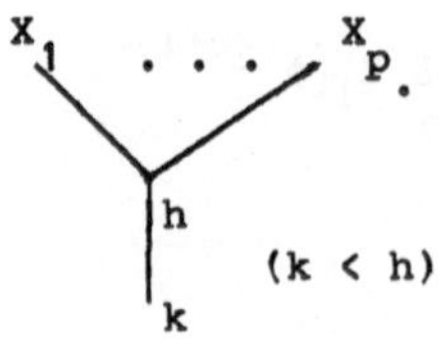

The succession of simple expectations $\xi_{(k,U]}(V) \equiv \xi_{k+1}\ldots\xi_U(V)$ is pictured as a segment ending in an <u>endpoint</u> V:

$$\xi_{(k,U)}(V) \leftrightarrow \quad (k<U). \qquad (2.12)$$

A tree τ is built out of the components (2.10)-(2.12). At the bottom of τ is the <u>root</u>, joined by a segment to the lowest fork F, and at the top of τ are endpoints. The number of endpoints is the order of perturbation theory. For example, a 10th order tree contributing to V_r^U is

$$(\tau,\vec{h}) = \qquad (2.13)$$

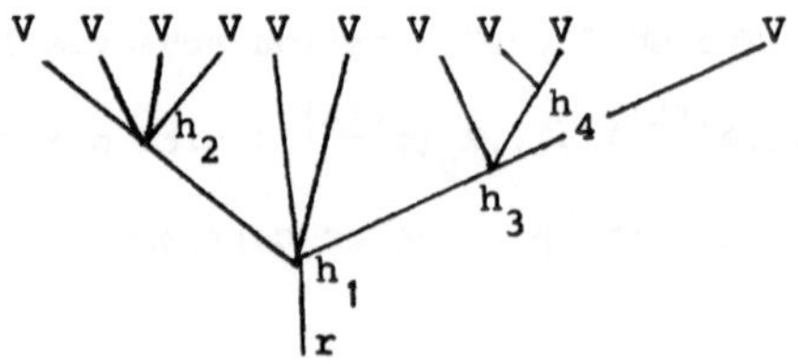

where $\vec{h} = (h_1, h_2, h_3, h_4)$ satisfies

$$r < h_1 < h_2 \leq U \qquad (2.14)$$
$$h_1 < h_3 < h_4 \leq U.$$

For convenience in referring to the example (2.13) we attach the name f_j to the fork with scale label h_j.

There is a natural partial ordering on the forks on a tree; namely, $f < f'$ if the fork f lies below f' on the tree. As in (2.14) the associated scales satisfy $h_f < h_{f'}$. All scales satisfy $r < h \leq U$ where the <u>root scale</u> r is determined by the scale of the external field $\Phi^{(\leq r)}$.

On the basis of the graphical description of V_k^U given after (2.9) we describe the graphs G associated with a tree $(\tau,\vec{h})$ as follows. If τ has n endpoints, G is a connected Feynman graph (with external legs) with n vertices, each corresponding to a monomial in V, and with connecting lines which may be hard $\left(C^{(h_i)}\right)$ or soft $\left(C^{(<h_i)}\right)$. These connections must be "consistent" with $(\tau,\vec{h})$. To define what we mean by "consistent" we consider G to be built up inductively from graphs corresponding to the forks of τ. Let G_f be the subgraph of G whose vertices correspond to endpoints of τ above f, whose internal lines are the lines of G joining these vertices, and whose external legs are the legs and remaining half-lines of G which are attached to these vertices.

The external legs of G_f correspond to fields $\Phi^{(<h_f)}$, and external legs of G to fields $\Phi^{(\underline{<}r)}$. If f' is a fork immediately above f, we write $\pi(f') = f$. Viewing, for each such f', the graph $G_{f'}$ as a "generalized vertex" for the graph G_f, we consider the <u>reduced graph</u>

$$g_f = G_f/\{G_{f'} : \pi(f') = f\} .$$

Here $G/\{G_1, \ldots, G_n\}$ is the graph obtained from G by contracting each subgraph G_i to a point. The lines of g_f are either hard $\left(C^{(h_f)}\right)$ or soft $\left(C^{(<h_f)}\right)$. We say G is <u>consistent with τ</u> if each g_f is connected by its hard lines. Note that the minimum number of hard lines that will satisfy this connectedness requirement is p_f-1, where p_f is the number of branches growing up from f. We denote the collection of graphs consistent with τ by $\mathcal{G}(\tau)$.

An example of a Feynman graph in QED corresponding to the tree (2.13) is

G = 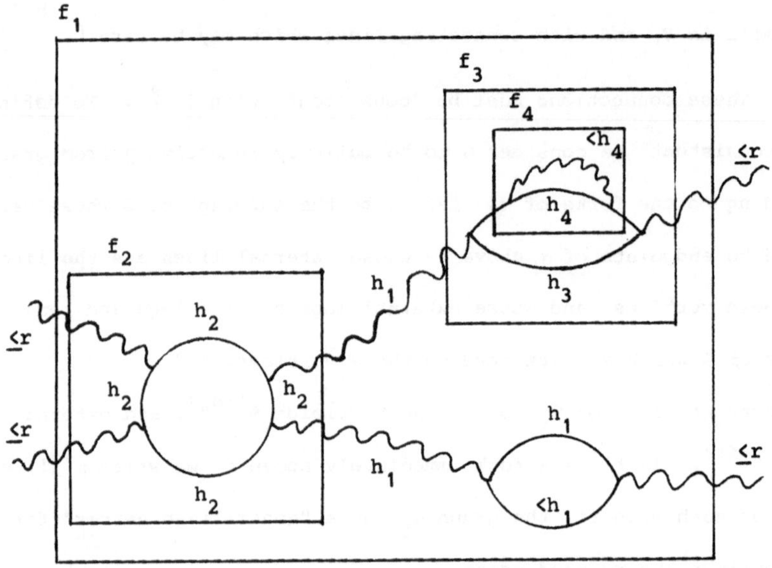

(2.15)

(In (2.15), the graph G_f associated with a fork f of τ is enclosed in a box labelled f. As usual, spinor lines are solid and photon lines wavy.)

Then g_{f_1} is given by

g_{f_1} = 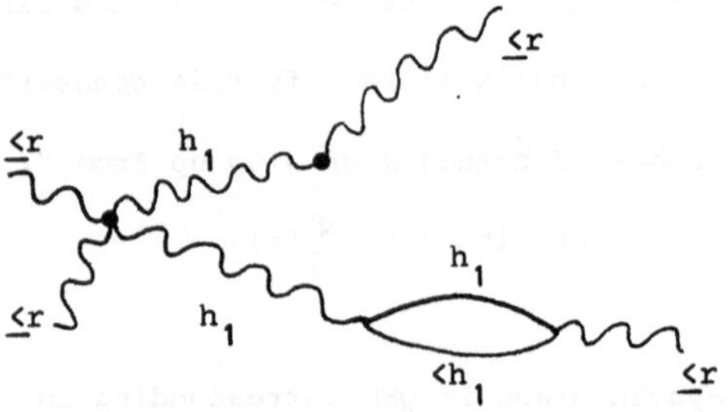

(2.16)

Inductively applying the evaluation of a graph for a single fork, we find the value $V^{\vec{h}}(G)$ of a graph G corresponding to $(\tau,\vec{h})$. It is the integral over the positions of the vertices of the product of the following factors:

(i) a Wick monomial in the external fields $:\prod_{\lambda\in\Lambda(G)} \Phi^{(\underline{<}r)}(x_\lambda):$;

(ii) for each f, a factor $C^{(h_f)}_{i_\ell j_\ell}(x_\ell,y_\ell)$ $\left(C^{(<h_f)}\right)$ for each hard (soft) line ℓ in g_f;

(iii) the usual combinatoric factor arising from the choices of legs to form lines and from hard and soft designations;

(iv) ± sign arising from commutations of fermi fields;

(v) coupling constants associated with each vertex.

We then define

$$V(\tau,\vec{h}) = \sum_{G\in\mathcal{G}(\tau)} V^{\vec{h}}(G). \tag{2.17}$$

The effective potential V^U_r is then the sum of the contributions $V(\tau,\vec{h})$ over τ and $\vec{h}$. For a fixed Feynman graph contributing to V^U_r, the finite number of trees which produce that graph give a decomposition of the graph which plays the joint role of the sum over "Hepp sectors"[4] and the sum over "Zimmermann forests"[5] in the BPHZ method.

There are two standard conventions about what we mean by "distinct" trees τ in this sum:

a) We consider the trees to be "plane" trees, i.e. for the purpose of determining whether two trees are the same they are assumed to be embedded in a plane: trees which differ by a reordering of distinct subtrees are distinct. With this convention we obtain

$$V^U_r = \sum_{\tau,\vec{h}} \frac{1}{n(\tau)} V(\tau,\vec{h}) \tag{2.18a}$$

where

$$n(\tau) = \prod_{f\in\mathfrak{F}(\tau)} p_f!$$

where p_f is the number of branches growing up from the fork f, and $\mathfrak{F}(\tau)$ is the set of forks of τ.

or b) We (inductively) identify τ and τ' if the subtrees of τ emerging from the first fork $F(\tau)$ can be identified in a 1-1 way with the subtrees of τ' emerging from $F(\tau')$; in other words, trees which differ by a reordering of subtrees at a fork are identified. Let $\beta_1,\ldots,\beta_b$ be the distinct subtrees of

τ emerging from $F(\tau)$, each occurring $n_1,\ldots,n_b$ times, where $n_1+\ldots+n_b = n$. Then, in the sum over $i_1,\ldots,i_n$ in (2.9), there are $\binom{n}{n_1\ldots n_b}$ equal terms which are identified with the tree τ. As a result,

$$V_r^U = \sum_{\tau,\vec{h}} \frac{1}{m(\tau)} V(\tau,\vec{h}) \tag{2.18b}$$

where the combinatoric factor $m(\tau)$ is given inductively by

$$\frac{1}{m(\tau)} = \frac{1}{n!}\binom{n}{n_1\ldots n_b} \prod_{j=1}^{b} \frac{1}{m(\beta_j)^{n_j}}$$

or

$$m(\tau) = \prod_{j=1}^{b} n_j!\ m(\beta_j)^{n_j}.$$

In this monograph we make the choice a).

In the tree expansion developed above we have chosen to Wick order the bare and effective potentials. We could, of course, have chosen not to do so. This would have resulted in only the following modifications to the above tree expansion:

(a) $\mathcal{E}_{(k,h)}$, which in (2.11) stands for $\int\cdot dP(\Phi^{(k+1)}) \ldots dP(\Phi^{(h-1)})$, would be replaced by evaluation at $\Phi^{(k+1)} = \ldots = \Phi^{(h-1)} = 0$.

(b) $\mathcal{E}_h^T(X_1)$, which in (2.11) is $\mathcal{E}_h(X_1)$, would be replaced by

$$\int X_1(\Phi^{(\leq h)})\, dP(\Phi^{(h)}) - X_1(\Phi^{(<h)}).$$

(c) There would be no soft lines in the Feynman graphs G.

Bounds on graphs in the unrenormalized expansion

We shall find bounds on the graphs contributing to each $V(\tau,\vec{h})$ which show in a natural way why renormalization is necessary and why it works.

Let G be a graph contributing a term $V(G)$ to $V(\tau,\vec{h})$. Then

$$V(G) = \int G(x_1,\ldots,x_s) :\Phi^{(\leq r)}(x_1)^{p_1} \ldots \Phi^{(\leq r)}(x_s)^{p_s}: dx_1\ldots dx_s$$

where we use the letter G both for the graph and its distributional kernel $G(x_1,\ldots,x_s)$, obtained by integrating the product of values attached to all lines and vertices of G with respect to internal vertex positions $x_{s+1},\ldots,x_n$. Here

a vertex of G is <u>external</u> if it has at least one external leg attached to it (otherwise the vertex is <u>internal</u>);

$\mathcal{V}(G)$ is the set of vertices of G;

$n = v(G) = |\mathcal{V}(G)|$ = the number of vertices of G;

s is the number of external vertices $v^e(G)$ of G;

$\ell(G) = |\mathcal{L}(G)|$ = the number of lines of G.

When discussing bounds on a graph we also use the symbol G to refer to a graph summed over the allowed $\vec{h}$ values in (2.18); i.e.,

$$r < h_f \leq U, \qquad h_f - h_{\pi(f)} > 0, \tag{2.19}$$

where $\pi(f)$ is the fork of τ immediately below f. If F is the lowest fork of τ we let $\pi(F)$ be the root of τ and we set $h_{\pi(F)} = 0$.

Our goal is to obtain bounds on the summed kernel that are uniform in the UV cutoff U.

It is natural to measure the size of G by some norm on the kernel like

$$\|G\|_0 \equiv \int |G(0,x_2,\ldots, x_s)|dx_2\ldots dx_s \quad \text{or} \quad \sup|G(x_1,\ldots,x_s)|. \tag{2.20}$$

The first norm amounts to taking the supremum of G in momentum space (after removing the delta function $\delta(p_1+\ldots+p_s)$). It works well for models containing only massive fields, but not for models with IR divergences where G fails to have integrable decay when the x_i's $\to \infty$ in some directions (equivalently, G has singularities at exceptional momenta). On the other hand, the second norm, while immune to a lack of decay as $x \to \infty$, is not suitable either. For, in general, UV convergent graphs have (integrable) local singularities in x-space corresponding to the lack of integrability at $p = \infty$ of propagators like $\frac{1}{p^2+m^2}$. We shall use a norm that is a combination of the two in (2.20):

$$\|G\| = \sup \left\{\left|\int dx_1\ldots dx_s\, G(x_1,\ldots,x_s) \prod_{i=1}^{s} f_i(x_i)\right| : f_i \in \mathcal{C}\right\} \tag{2.21a}$$

where

$$\mathcal{C} = \bigcup_{x\in\mathbf{R}^d} \left\{f\in C_0^\infty\big(B_1(x)\big) : \sup|\partial^m f(y)| \leq 1,\ |m| \leq d\right\} \tag{2.21b}$$

and $B_1(x) = \{y: \|x-y\| \leq 1\}$. Hence, each function in $\mathcal{C}$ is infinitely differentiable, bounded by 1 together with its derivatives of order $\leq$ d and supported in some sphere of radius 1. If $\|G\|$ is bounded uniformly in U we say that the graph G is <u>convergent</u>; otherwise we say that G is <u>divergent</u>. If G is convergent, then the kernel $\lim_{U\to\infty} G$ exists and must be locally integrable, but can have mild local singularities, and must be bounded, but not necessarily integrable, at ∞; it is certainly a tempered distribution.

To find a bound on $\|G\|$, we use the following bounds on hard and soft lines:

$$|C_\ell^{(h_\ell)}(x,y)| \leq c\, M^{d_\ell h_\ell} e^{-M^{h_\ell}|x-y|}$$

$$|C_\ell^{(<h_\ell)}(x,y)| \leq c\, M^{d_\ell h_\ell}. \tag{2.22}$$

See (1.3); here $d_\ell = 2\delta_i$ where Φ_i is the field associated to the line ℓ, and h_ℓ is the scale of the fork at which the line ℓ first appears, i.e., $\ell \in \mathcal{L}(g_f)$. Since all constants associated with lines or vertices can clearly be accommodated in a factor of the form $K^{\ell(G)}$, we deduce that

$$|G(x_1,\ldots,x_s)| \leq K^{\ell(G)} \prod_{\ell\in\mathcal{L}(G)} M^{d_\ell h_\ell} \int dx_{s+1}\ldots dx_n \prod_{\ell\in\mathcal{M}} e^{-M^{h_\ell}|\ell|} \tag{2.23}$$

where $|\ell| = |x_\ell - y_\ell|$ is the length of ℓ, and $\mathcal{M}$ is a minimal set of hard lines that connect each g_f for $f\in\mathcal{F}(\tau)$, the set of forks of τ. Note that:

(1) $\mathcal{M}$ contains $v(G)-1$ lines.

(2) If $f_1,\ldots,f_{p_f}$ are the forks (or endpoints) immediately above f, then $G_{f_1},\ldots,G_{f_{p_f}}$ are simply connected by p_f-1 lines of $\mathcal{M}$ of frequency h_f.

(3) $$\sum_{f\in\mathcal{F}(\tau)} (p_f-1) = v(G)-1. \tag{2.24}$$

From (2.21) and (2.23), we have, for all choices of f_i in $\mathcal{C}$,

$$\|G\| \leq K^{\ell(G)} \prod_{\ell\in\mathcal{L}(G)} M^{d_\ell h_\ell} \int \prod_{i=1}^{n} dx_i \prod_{\ell\in\mathcal{M}} e^{-M^{h_\ell}|\ell|} \prod_{i=1}^{s} |f_i(x_i)| . \tag{2.25}$$

To bound the integral in (2.25), we perform the integration over the vertices

x_i in the following order. One of the vertices, say x_1, is external. It is integrated last. We drop the factors $|f_i(x_i)| \le 1$ for $i \ne 1$ from (2.25). Now the lines of $\mathcal{M}$ form a tree graph T connecting all the x_i's. We first integrate the x_i's $(i \ne 1)$ which are endpoints of T and we erase the line attached to each of them. We next integrate the new endpoints $(\ne x_1)$ and erase their attached lines, etc. Associating each vertex $\ne x_1$ with its attached line, $x_{i(\ell)} \leftrightarrow \ell$, we obtain from the integration of $x_{i(\ell)}$ the factor

$$\int dx_{i(\ell)}\, e^{-M^{h_\ell}|x_{i(\ell)}-y|} = c\, M^{-h_\ell d}. \tag{2.26}$$

Overall we obtain a factor $c^{n-1} \prod_{\ell \in \mathcal{M}} M^{-h_\ell d}$. For the last vertex, we have

$$\int dx_1 |f_1(x)| \le K. \tag{2.27}$$

Substituting into (2.25), we have (absorbing the constants into $K^{\ell(G)}$)

$$\|G\| \le K^{\ell(G)} \prod_{\ell \in \mathcal{L}(G)} M^{d_\ell h_\ell} \prod_{\ell \in \mathcal{M}} M^{-h_\ell d} = K^{\ell(G)} \prod_{\ell \in \mathcal{L}(G)} M^{d_\ell h_\ell} \prod_{f \in \mathcal{F}(\tau)} M^{-h_f d(p_f - 1)} \tag{2.28}$$

by property (2) of $\mathcal{M}$.

We next rewrite (2.28) in a form involving differences of scales $h_f - h_{\pi(f)}$ by means of the following elementary lemma.

<u>Lemma 2.1</u>. (Summation by parts on a tree) Let τ be a tree with functions a_f, b_f defined on the forks f of τ. For f_1 a fixed fork or the root of τ,

$$\sum_{f > f_1} (a_f - a_{f_1}) b_f = \sum_{f > f_1} (a_f - a_{\pi(f)}) B_f \tag{2.29}$$

where $B_f = \sum_{f' \ge f} b_{f'}$.

<u>Proof</u>. $\text{R.S.} = \sum_{f > f_1} (a_f - a_{\pi(f)}) \sum_{f' \ge f} b_{f'}$

$$= \sum_{f' > f_1} b_{f'} \sum_{f_1 < f \le f'} (a_f - a_{\pi(f)}).$$

The summation over f takes place over a totally ordered set and so telescopes to

$a_{f'} - a_{f_1}$. ∎

For f_1 the root of τ and for $a_f = h_f$ (recall that $h_{f_1} = 0$), (2.29) reads

$$\sum_{f\in\mathfrak{F}(\tau)} h_f b_f = \sum_{f\in\mathfrak{F}(\tau)} \left(h_f - h_{\pi(f)}\right) B_f. \tag{2.30}$$

We apply (2.30) with

(1) $b_f = v(g_f) - 1 = p_f - 1$ so that $B_f = v(G_f) - 1$,

(2) $b_f = \sum_{\ell\in\mathcal{L}(g_f)} d_\ell$ so that $B_f = \sum_{\ell\in\mathcal{L}(G_f)} d_\ell$

to obtain

$$\sum_f h_f(p_f - 1) = \sum_f \left(h_f - h_{\pi(f)}\right)\left(v(G_f) - 1\right) \tag{2.31}$$

and

$$\sum_{\ell\in\mathcal{L}(G)} h_\ell d_\ell = \sum_f h_f \sum_{\ell\in\mathcal{L}(g_f)} d_\ell = \sum_f \left(h_f - h_{\pi(f)}\right) \sum_{\ell\in\mathcal{L}(G_f)} d_\ell. \tag{2.32}$$

By (2.31) and (2.32) the exponent in (2.28) is

$$\sum_{\ell\in\mathcal{L}(G)} h_\ell d_\ell - d\sum_f h_f(p_f - 1) = \sum_f \left(h_f - h_{\pi(f)}\right) D(G_f) \tag{2.33}$$

where $D(H)$ is the usual <u>UV degree of divergence</u> of a graph H:

$$D(H) \equiv \sum_{\ell\in\mathcal{L}(H)} d_\ell - d\left(v(H) - 1\right). \tag{2.34}$$

Combining (2.28) and (2.33) we obtain the basic bound on unrenormalized graphs:

<u>Theorem 2.2</u>. If G is a graph contributing to some $V(\tau,\vec{h})$ in (2.17),

$$\|G\| \le K^{\ell(G)} \prod_{f\in\mathfrak{F}(\tau)} M^{\left(h_f - h_{\pi(f)}\right)D(G_f)}, \tag{2.35}$$

where $h_{\pi(F)} = 0$ for the lowest fork F of τ.

Theorem 2.2 provides the basic power counting criterion for the convergence of an unrenormalized graph G. To prove the existence of the unrenormalized effective potential V_r^U in perturbation theory, we would have to show that the sum (2.18a) converges uniformly in U. For a specific Feynman graph contributing to V_r^U, there is a finite number of trees which produce that graph. Hence the sum over τ in (2.18a) is finite. The $\vec{h}$ sum in (2.18a) takes place over the set

(2.19).

Clearly, the sum of (2.35) over these $\vec{h}$'s converges uniformly in U if and only if each $D(G_f) < 0$. When $D(G_f) \geq 0$ we must renormalize the subgraph G_f.

<u>Example</u>(QED_4) A subgraph G_f with n vertices, ℓ_p photon lines, λ_p external photon legs, ℓ_s spinor lines and λ_s external spinor legs has

$$\begin{aligned} D(G_f) &= 2\ell_p + 3\ell_s - 4(n-1) \\ &= 2(n-\lambda_p)/2 + 3(2n-\lambda_s)/2 - 4n + 4 \\ &= 4 - \lambda_p - \frac{3}{2}\lambda_s. \end{aligned} \tag{2.36}$$

Since λ_s is even and λ_p is even if $\lambda_s = 0$ (Furry's Theorem, Lemma 3.3), the subgraphs requiring renormalization are those with

$$(\lambda_p,\lambda_s) = (0,2),\ (1,2),\ (2,0) \text{ or } (4,0).$$

<u>The renormalized expansion</u>

We now attach the subscript "un" to the unrenormalized effective potential discussed above and we use the symbol V_k^U for the renormalized effective potential.

In the renormalized expansion we modify the bare potential $V = V_{U,un}^U$ by adding appropriate counterterms

$$\delta V^U = -\int \delta\mathcal{L}^U\, dx = -\int \Sigma\, \lambda_j P_j\left(\Phi^{(\leq U)}\right) dx$$

where λ_j depends on U and $\delta\mathcal{L}^U$ may include all possible Euclidean invariant, local terms of dimension $\leq d$, including terms from the free Lagrangian. For example, in QED_4,

$$\begin{aligned} \delta\mathcal{L}^U = \lambda_1 :\bar\psi^{(\leq U)}\psi^{(\leq U)}: + \lambda_2 :\bar\psi^{(\leq U)}(-i\not\partial)\psi^{(\leq U)}: + \lambda_3 :\bar\psi^{(\leq U)}\not A^{(U)}\psi^{(\leq U)}: \\ + \lambda_4 :\left(F^{(\leq U)}\right)^2: + \lambda_5 :\left(\partial\cdot A^{(\leq U)}\right)^2: + \lambda_6 :\left(A^{(\leq U)}\right)^2: + \lambda_7 :\left(A^{(\leq U)}\right)^4:, \end{aligned} \tag{2.37}$$

where $\dim \psi = 3/2$, $\dim A = 1$, $\dim F = 2$, $\dim \partial = 1$. Spinor and vector indices have been suppressed and are to be summed over. Note that the terms $\partial\cdot A$,

$(\partial.\partial)\partial.A$, and $(\partial.A)A^2$ have dimension ≤ 4 but are absent by Furry's Theorem (Lemma 3.3) and that, as a total derivative, $(\partial.\partial)A^2$ is absent.

Each coefficient λ_j is of the form $\lambda_j = \sum_{n=2}^{\infty} \lambda_j^{(n)}$, where $\lambda_j^{(n)}$ is of order n in e, and must be chosen in such a way that there are no divergences to that order of perturbation theory in the effective potential V_r^U. The tree expansion provides an inductive method for making a choice of counterterms.

We first consider how to choose the counterterms corresponding to a tree with a single fork f. We shall select a local term LG_f corresponding to each divergent graph G_f, such that $G_{f,R} = (1-L)G_f$ is convergent.

For example in QED_4, consider a graph G_f with two external spinor legs that contributes to the tree

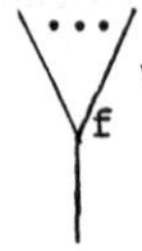

with n endpoints. G_f is of order n in e, $D(G_f) = 1$ (see (2.36)), and

$$V(G_f) = \int :\bar{\psi}^{(\leq r)}(x)G_f(x,y)\psi^{(\leq r)}(y):\, dxdy$$

for some kernel

$$G_f(x,y) = \sum_{h_f=r+1}^{U} G_f^{(h_f)}(x,y). \tag{2.38}$$

We renormalize G_f by performing a Taylor subtraction on $\psi^{(\leq r)}(y)$ to obtain

$$V(G_{f,R}) \equiv \int :\bar{\psi}^{(\leq r)}(x)G_f(x,y)\left[\psi^{(\leq r)}(y)-\psi^{(\leq r)}(x)-(y-x)^{\mu}\partial_{\mu}\psi^{(\leq r)}(x)\right]:dxdy$$
$$\equiv V\left((1-L)G_f\right). \tag{2.39}$$

Note that the subtracted term LG_f is local in $\bar{\psi}$, ψ (we suppress the superscript $(\leq r)$):

$$V(LG_f) = \int :\bar{\psi}(x)\left(\int G_f(x,y)dy\right)\psi(x):\, dx$$
$$+ \int :\bar{\psi}(x)\left(\int G_f(x,y)(y-x)^{\mu}dy\right)\partial_{\mu}\psi(x):\, dx.$$

Now by Euclidean invariance,

$$c_0 = \int G_f(x,y)dy \quad \text{and} \quad c_1^\mu = i\int G_f(x,y)(y-x)^\mu dy \tag{2.40a}$$

are independent of x, and $c_1^\mu = c_1\gamma^\mu$ with c_0 and c_1 independent of spinor indices. Hence

$$V(LG_f) = c_0 \int :\bar{\psi}\psi:dx + c_1 \int :\bar{\psi}(-i\not{\partial})\psi:dx. \tag{2.40b}$$

The renormalized kernel in (2.39) may be written

$$G_{f,R}(x,y) = \int dy' G_f(x,y')\left[\delta(y-y') - \delta(y-x) - \delta(y-x)(y'-x)^\mu \frac{\partial}{\partial y^\mu}\right]$$

and so by the definition (2.21)

$$\|G_{f,R}\| = \sup_{f_i \in \mathcal{C}} \left|\int dxdy\; f_1(x)G_f(x,y)\left[f_2(y)-f_2(x)-(y-x)^\mu\partial_\mu f_2(x)\right]\right|. \tag{2.41}$$

Now by Taylor's Theorem,

$$f_2(y)-f_2(x)-(y-x)^\mu\partial_\mu f_2(x) = \frac{1}{2}(y-x)^\mu(y-x)^\nu\partial_\mu\partial_\nu f_2\big(sy+(1-s)x\big) \tag{2.42}$$

for some s, $0 \leq s \leq 1$. Since $|\partial_\mu\partial_\nu f_2| \leq 1$, we see from (2.41) and (2.42) that

$$\|G_{f,R}\| \leq \frac{1}{2} \sup_\xi \int_{B_1(\xi)} dx \int dy \;|G_f(x,y)|\; |y-x|^2. \tag{2.43}$$

We have already estimated this integral without the factor $|y-x|^2$ in (2.28). By a scaling argument we see that this factor supplies an additional factor M^{-2h_f}. The contribution $M^{h_f D(G_f)}$ to the bound (2.35) on $\|G_f\|$ is therefore modified by (2.44) to $M^{h_f D(G_{f,R})}$ where

$$D(G_{f,R}) = D(G_f) - 2 = -1.$$

We conclude that $G_{f,R} = (1-L)G_f$ is convergent.

The subtraction LG_f can be implemented by an appropriate contribution to δV^U. However, note that, by (2.40a) and (2.38), $c_i = \sum_{h=r+1}^{U} c_i^{(h)}$; hence c_i depends on r whereas <u>δV^U is not allowed to depend on r</u>. The resolution of this conflict is to introduce the missing scales $c_i' = \sum_{h=0}^{r} c_i^{(h)}$, and in (2.37) to choose

$$\lambda_i^{(n)} = c_i + c_i' + \text{contributions from other graphs.} \tag{2.44}$$

Then $\lambda_i^{(n)}$ is independent of r, the contribution c_i effects the cancellation (2.39), while the extra term c_i' introduces a "useless counterterm" which has to be treated as an extra interaction vertex. Since c_i' is independent of U the demonstration of UV convergence given above is not spoiled, but the dependence of c_i' on r requires further attention.

In the above example, the local part LG_f was taken to be the projection of the monomial $V(G_f)$ onto the span of the local monomials with dimension $\leq$ d. This illustrates the general renormalization prescription for strictly renormalizable models: a graph G is renormalized if and only its <u>external degree of divergence</u> $\delta(G)$ is nonnegative, where

$$\delta(G) \equiv d - \sum_{\lambda\in\Lambda(G)} \delta_\lambda . \tag{2.45}$$

In general, we will be faced with renormalizing monomials of the form

$$G = \int G(x_1,\ldots,x_s)\, \Pi(x_1,\ldots,x_s)\, dx_1\ldots dx_s \tag{2.46}$$

where

$$\Pi(x_1,\ldots,x_s) = :\partial^{q_1}\Phi_{i_1}(x_1)\ \ldots\ \partial^{q_s}\Phi_{i_s}(x_s): . \tag{2.47}$$

The dimension of Π is $\Delta \equiv q_1 + \ldots + q_s + \delta_{i_1} + \ldots + \delta_{i_s}$ (here $\delta_i = \frac{d-2}{2}$ for a bose field and $\frac{d-1}{2}$ for a fermi field). The external degree of G is $\delta \equiv \delta(G) = d - \Delta$. We define

$$LG = \int G(x_1,\ldots,x_s)\, (L\Pi)(x_1,\ldots,x_s)\, dx_1\ldots dx_s \tag{2.48a}$$

where $L\Pi = 0$ if $\delta < 0$ and $L\Pi$ is the Taylor polynomial approximation of order δ for Π about x_1 if $\delta \geq 0$; i.e.,

$$(L\Pi)(x_1,\ldots,x_s) = \sum_{j=0}^{\delta} \frac{1}{j!}\left(\frac{d}{dt}\right)^j \Pi(x_1,x_2(t),\ldots,x_s(t))\Big|_{t=0} \tag{2.48b}$$

where $x_j(t) = x_1 + t(x_j - x_1)$. For example, in QED_4,

$$L\int :\bar\psi(x_1)G(x_1,x_2)\psi(x_2):dx_1dx_2 = \int dx_1 :\bar\psi(x_1)\left(\int G(0,x_2)dx_2\right)\psi(x_1):$$

$$+ \int dx_1 :\bar\psi(x_1)\left(\int dx_2 x_2^\mu G(0,x_2)\right)\partial_\mu\psi(x_1).$$

By the translation invariance of $G(x_1,\ldots,x_s)$, the value of LG does not depend on which coordinate x_j we single out as the centre of the Taylor polynomial approximation.

As in (2.39) a renormalized graph is defined to be

$$RG = (1 - L)G = \int G(x_1,\ldots,x_s)[(\Pi - L\Pi)(x_1,\ldots,x_s)]dx_1\ldots dx_s .$$

By Taylor's Theorem we can write

$$RG = \frac{1}{\delta!}\int_0^1 dt(1-t)^{\delta}\left(\frac{d}{dt}\right)^{\delta+1}\int G(x_1,\ldots,x_s)\Pi\big(x_1,x_2(t),\ldots,x_s(t)\big)dx_1\ldots dx_s . \qquad (2.49)$$

The tree notation can be modified to describe these renormalized expressions by appending an extra label on each fork as follows. We write

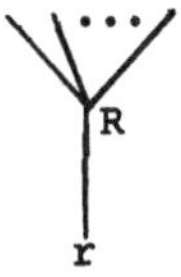 $= \sum_{h_f=r+1}^{U}$ 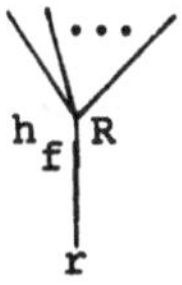 $= \sum_{h_f=r+1}^{U} (1-L)$ 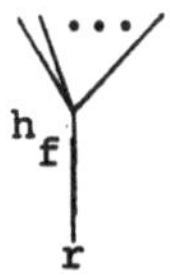 ,

and, for the contribution to V_r^U from the extra unused counterterms, we write

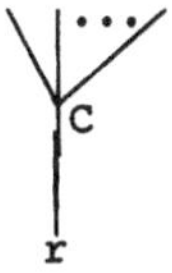 $= \sum_{h_f=0}^{r}$ 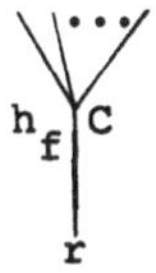 $= - \sum_{h_f=0}^{r} L$ 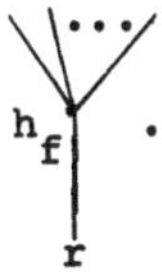 .

Here L is defined on each graph contributing to the tree 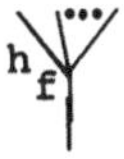 as in (2.48) except that the external fields are $\phi^{(\leq r)}$ (even when $h_f \leq r$ as in the C-fork above).

A general <u>renormalized tree</u>, i.e. a tree that occurs in the renormalized expansion, has a label ρ_f = R or C at each fork f. We write $\vec{\rho}$ for $(\rho_f)_{f\in\mathcal{F}(\tau)}$. As in the unrenormalized expansion,

$$h_{\pi(f)} < h_f \leq U \quad \text{if } \rho_f = R$$

but (2.50)

$$0 \leq h_f \leq h_{\pi(f)} \quad \text{if } \rho_f = C.$$

We denote the set of scales (2.50) by $\mathcal{H}(\tau,\vec{\rho})$. As before, each endpoint of τ is associated with a monomial from V, and the tree is evaluated by applying the following rules: If $X_1,\ldots,X_p$ are Wick monomials arising from V, or from an R-fork or C-fork, then

 $= X_1(\Phi^{(\leq k)})$

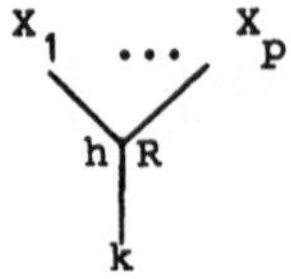 $= \chi(h>k)\ \mathcal{E}_{(k,h)}(1-L)\mathcal{E}_h^T\big(X_1(\Phi^{(\leq h)}),\ldots,X_p(\Phi^{(\leq h)})\big)$ (2.51)

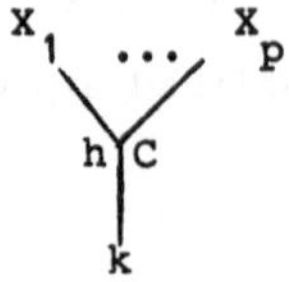 $= -\chi(h\leq k)\ L\mathcal{E}_h^T\big(X_1(\Phi^{(\leq h)}),\ldots,X_p(\Phi^{(\leq h)})\big)\Big|_{\Phi^{(\leq h-1)} \text{ replaced by } \Phi^{(\leq k)}.}$ (2.52)

Thus a C-fork feeds into the fork $\pi(f)$ as a sum of local "interaction" monomials in $\Phi^{(\leq k)}$ where $k = h_{\pi(f)}$, with a coefficient depending on k via the sum $\sum_{h=0}^{k}$.

We introduce the notations

 ,  and 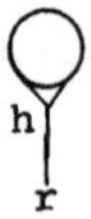

to denote the sum of the values of all non-trivial trees whose lowest fork F and whose root have the labels shown. Here, a <u>non-trivial</u> tree is one with at least one fork; each fork $f > F$ has either an R or C label; the sum is over all possible labelled trees, all choices of a monomial from V for each endpoint, and all scales h_f $(f > F)$ satisfying (2.50); and, the sum includes an appropriate combinatoric factor for each labelled tree, as in (2.18). In these sums we omit constant terms, that is terms independent of the field $\Phi^{(\leq r)}$. It will also be convenient to write

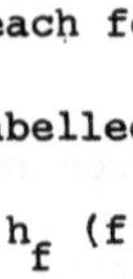 $= \sum_{h=r+1}^{U}$ 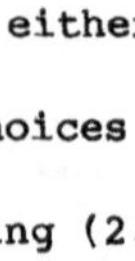 and 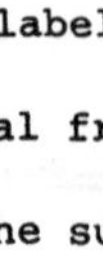 $= \sum_{h=0}^{r}$ 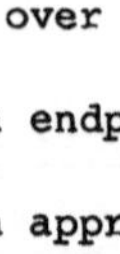 .

We then have the basic identity

$\sum_{h=r+1}^{U}$ 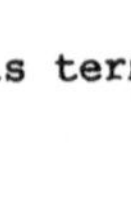 $- \sum_{h=0}^{U}$ 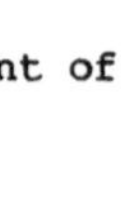 $=$ 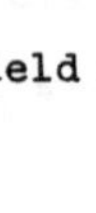 $+$  (2.53)

Motivated by the example (2.46) we choose the counterterms to be

$\delta V^U =$  (2.54)

so that the renormalized effective potential at scale U is

$V_U^U = V_{U,un}^U + \delta V^U =$ 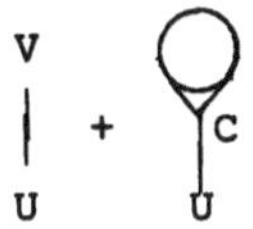 . (2.55)

Then we have:

<u>Theorem 2.3.</u> (<u>Renormalized tree expansion</u>) For $0 \leq r \leq U$

$V_r^U =$ 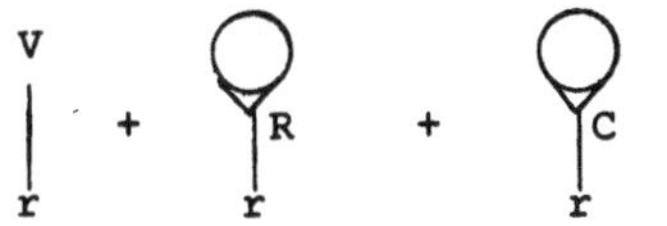 .

$$= V\left(\Phi^{(\leq r)}\right) + \sum_{\tau} \frac{1}{n(\tau)} \sum_{\vec{\rho}} \sum_{\vec{h} \in \mathcal{H}(\tau,\vec{\rho})} V(\tau,\vec{\rho},\vec{h}), \tag{2.56}$$

where the first sum is over all non-trivial trees τ, $n(\tau)$ is given in (2.18a), and the value $V(\tau,\rho,\vec{h})$ is determined by the rules (2.51)-(2.52) for R and C labels.

<u>Proof.</u> For $r = U$, the middle term in (2.56) is zero and so (2.56) is true by definition $\left((2.55)\right)$. Suppose (2.56) holds for a fixed $r \leq U$. Then integrating out $\Phi^{(r)}$, we have by (2.9)

$V_{r-1}^U =$ 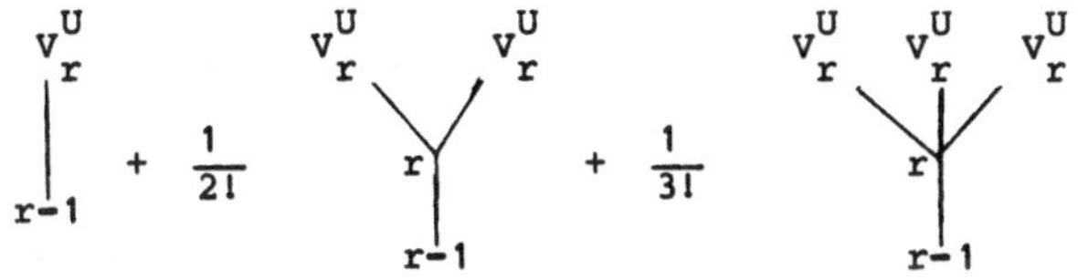 $+ \ldots$ (2.57)

By the inductive hypothesis (2.56) and the identity (2.53) the first term in (2.57) is given by

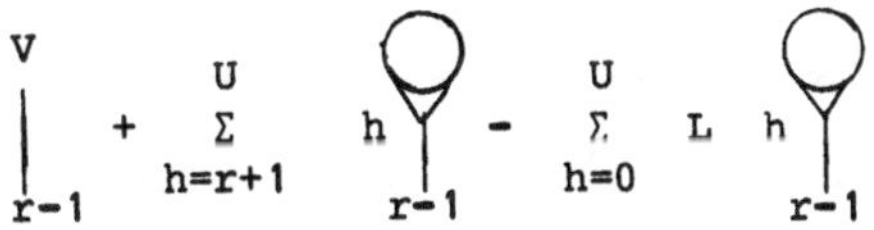 (2.58)

while the remaining terms combine to give  . Hence

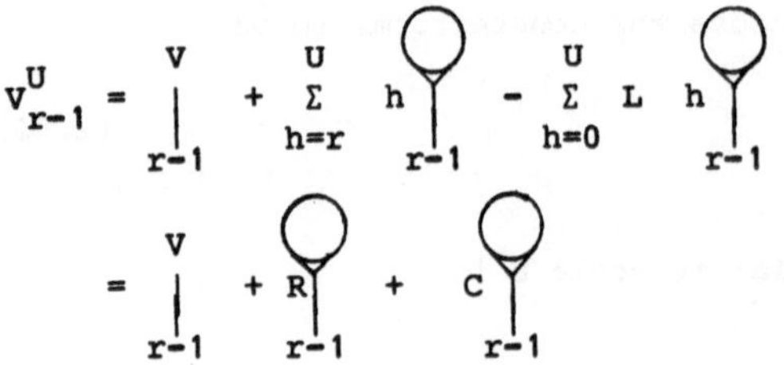

by (2.53). This establishes the inductive hypothesis for r-1. ∎

As r decreases from U to 0 in (2.56), the counterterms 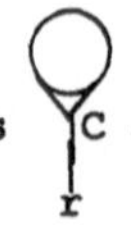 are gradually used up, usefully (to renormalize trees) and uselessly (to give C forks in the trees). When finally we integrate out $\Phi^{(0)}$ in (2.56), leaving only the external field Φ^{e} (see (1.5)), we obtain, as in the proof of the theorem,

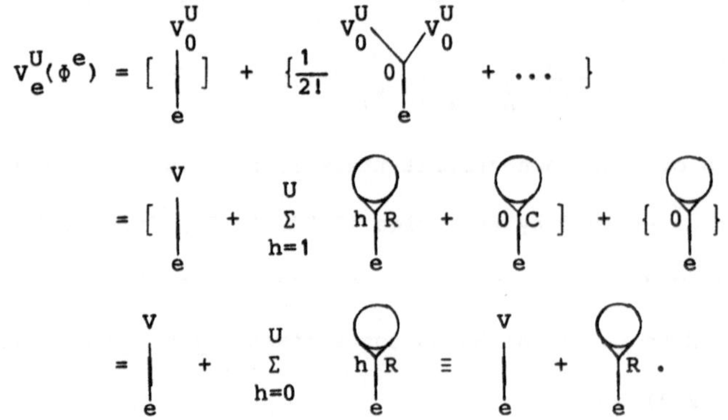

(2.59)

The counterterms have now been completely used up. (2.59) may be viewed as equation (2.56) with $r = -1$.

Now the counterterms δV^U are introduced in the GN formalism as a sum (2.54) over trees. One might ask whether the GN definition of counterterms agrees with a prescription like BPHZ, in which graphs are not decomposed as a sum over trees. To see that the answer is yes, let δV_n^U be the sum of all terms in (2.54) which are of order n in the coupling constant (where we assume for simplicity that there is one coupling constant λ); given a sum of terms Σ let $L_n\Sigma$ be those terms in $L\Sigma$ which are of order n in λ. We also write $\delta V_{\leq n}^U = \sum_{j=2}^{n} \delta V_j^U$, etc.

<u>Corollary 2.4</u>. For $n \geq 1$,

$$\delta V^{U}_{n+1}(\phi^{e}) = -L_{n+1} \log \int e^{(V+\delta V^{U}_{\leq n})\left(\phi^{e}+\phi^{(\leq U)}\right)} dP\left(\phi^{(\leq U)}\right). \tag{2.60}$$

<u>Proof</u>. Applying L_{n+1} to (2.59) gives

$L_{n+1}\, V^{U}_{e} = L_{n+1}$  $+\; L_{n+1}$  .

The first term on the right vanishes since V is first order in λ, and the second term vanishes since L is a projection operator and  contains a factor $(1-L)$ (see (2.52) and note that L commutes with $\mathcal{E}_{(k,h)}$). Hence

$$L_{n+1}\, V^{U}_{e} = 0. \tag{2.61}$$

Now

$$L_{n+1}\, V^{U}_{e} = L_{n+1} \log \int e^{V+\delta V^{U}_{\leq n+1}} dP\left(\phi^{(\leq U)}\right)$$

since $\delta V^{U}_{>n+1}$ does not contribute to order n+1. Moreover δV^{U}_{n+1} contributes only the trivial tree, and so

$$L_{n+1}\, V^{U}_{e} = \delta V^{U}_{n+1} + L_{n+1} \log \int e^{V+\delta V^{U}_{\leq n}} dP\left(\phi^{(\leq U)}\right). \tag{2.62}$$

The result follows from (2.61) and (2.62). ∎

Thus the GN counterterms, when summed up over all trees, agree with the usual counterterms defined inductively as in (2.60) with no reference to trees. The advantage of (2.60) is that invariance properties of the counterterms are manifest, whereas the advantage of the GN definition is that the cancellations in each "sector" (i.e. tree) are manifest.

Note that (2.61) in combination with (1.4) specifies our renormalization prescription. In the case of the ϕ^4_4 model, they say that the renormalized coupling constant (defined to be minus the four point connected Green's function, amputated by C^{-1} and evaluated at zero external momentum) is λ. Similar statements apply to the renormalized mass and field strength renormalization.

We write each monomial that contributes to the (unrenormalized) output from a fork f as

$$\int K_f(x_f)\, \Pi_f(x_f) dx_f \tag{2.63}$$

where the kernel K_f includes δ-functions for coinciding arguments, $x_f = (x_f^1, \ldots, x_f^n) \in R^{dn}$, and

$$\Pi_f(x_f) = :\partial^{q_1} \Phi_{i_1}^{(\leq k)}(x_f^1) \ldots \partial^{q_n} \Phi_{i_n}^{(\leq k)}(x_f^n): \tag{2.64}$$

where $k = h_{\pi(f)}$ and the x-derivatives ∂^{q_j} are "renormalization derivatives", i.e. derivatives which have been introduced by the Taylor operations (i.e. R- and C-operations) of the renormalized tree expansion, and not "interaction derivatives", i.e. ones present on the fields in the original interaction (the latter we include in the notation Φ_{i_j}).

We mention in passing that by integration by parts the x-derivatives in (2.63) may be placed either on the kernel or on the Wick monomial. Although this changes the apparent dimension of the monomial, subsequent R- and C-operations are not affected. To see this note that it follows from the definition (2.48) of L that

$$\partial_{x^j} L\, \Pi(x) = L\, \partial_{x^j} \Pi(x), \quad j \neq j_0, \tag{2.65}$$

where $x^o = x^{j_0}$ is the coordinate about which L localizes. When $j = j_0$, (2.65) fails as is, but it does hold when integrated against a translation invariant kernel K:

$$\int K(x)\, \partial_{x^j} L\, \Pi(x)\, dx = \int K(x)\, L\, \partial_{x^j} \Pi(x)\, dx. \tag{2.66}$$

(2.66) follows from (2.65) and integration by parts. In general in our formalism, the derivatives arise naturally on the fields and we leave them there without integrating by parts.

Consider the action of R^{δ_f} (see 2.49) on the monomial (2.63) where $\delta_f = d-\Delta_f \geq 0$, Δ_f being the dimension of Π_f (defined before (2.48)). Suppressing subscripts f, we have

$$R^{\delta} \int K(x)\, \Pi(x)\, dx = \frac{1}{\delta!} \int_0^1 dt (1-t)^{\delta}\, \partial_t^{\delta+1} \int K(x)\, \Pi\big(x(t)\big)\, dx \qquad (2.67)$$

where for $0 \leq t \leq 1$

$$x^j(t) = x^o + t(x^j - x^o).$$

(x^o, the centre of localization, is selected arbitrarily from among the arguments $x^1,\ldots,x^n$.) Now

$$\partial_t^{\delta+1}\, \Pi\big(x(t)\big) = (\Delta.\partial)^{\delta+1}\, \Pi\big(x(t)\big)$$

where $\Delta^j = x^j - x^o$ is not acted upon by $\partial^j = \frac{\partial}{\partial x^j}$. The change of variables $y = x(t)$, $dx = t^{-d(n-1)} dy$, $y^o = x^o$, $x = y(t^{-1})$ in (2.67) yields

$$R^{\delta} \int K(x)\, \Pi(x)\, dx = \int R^{\delta} K(y)\, \partial_y^{\delta+1}\, \Pi(y) dy \qquad (2.68)$$

where

$$R^{\delta} K(y) = \frac{1}{\delta!} \int_0^1 dt\ (1-t)^{\delta}\ t^{-d(n-1)} (\Delta^{\delta+1} K)\big(y(t^{-1})\big) \qquad (2.69)$$

$$(\Delta^{\delta+1} K)(x) = \prod_{i=1}^{\delta+1} (x^{j_i} - x^o) K(x) \qquad (2.70)$$

and

$$\partial_y^{\delta+1} = \prod_{i=1}^{\delta+1} \partial_{y^{j_i}},$$

the suppressed indices $j_1, \ldots, j_{\delta+1}$ in (2.68) being summed over. If $\delta<0$ we interpret $R^{\delta} K = K$ and $\partial_y^{\delta+1} = 1$.

At a C-fork f with degree $\delta \geq 0$, we have the output

$$- \int dx\, K_f(x) \sum_{m=0}^{\delta} \frac{1}{m!}\, \partial_t^m\, \Pi_f\big(x(t)\big)\Big|_{t=0} = - \int dx\, K_f(x) \sum_{m=0}^{\delta} \frac{1}{m!} (\Delta\cdot\partial_x)^m\, \Pi_f\big(x(0)\big)$$
$$= \sum_{m=0}^{\delta} \int dy\, (C_m K_f)(y)\, \partial^m\, \Pi_f(y) \qquad (2.71)$$

where

$$C_m K_f(y) = - \frac{1}{m!} \int \Delta^m\, K_f(y)\, \delta(y^o) dy \prod_{j \neq j_0} \delta(y^j - y^o)\ . \qquad (2.72)$$

We insert the expressions (2.68) and (2.71) for the Taylor operations into the tree expansion (2.56) and ask how nested Taylor operations are estimated. But first we ask what we mean by a "renormalized graph" in the expansion.

To fix the ideas consider the example of the tree

$\tau =$ 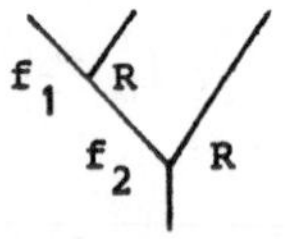

in the ϕ_4^4 model. There are many unrenormalized graphs associated with τ, determined according to which legs contract and whether they form hard or soft lines; but consider the single (unrenormalized) graph

$G_{f_2} =$ 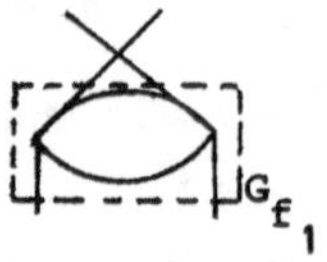

formed by first contracting 2 pairs of legs at f_1 to form the subgraph G_{f_1} (with 2 hard lines $C^{(h_1)}$) and then contracting 2 legs of G_{f_1} with 2 legs of a third vertex to give two hard lines $C^{(h_2)}$.

Now $\delta_{f_1} = D(G_{f_1}) = 0$ and so the renormalization of G_{f_1} involves subtracting a counterterm of degree 0 and RG_{f_1} one derivative on its external lines. Writing out the suppressed sum in (2.68) schematically we have

$RG_{f_1} =$ 

where we have chosen x^1 as the centre of localization so that a difference $\Delta x^j = x^j - x^1$ occurs for $j = 3, 4$ and the derivative acts either on $\phi(x_3)$ or $\phi(x_4)$ (denoted by a slash on the corresponding leg). K^j denotes the kernel (2.69) with $\Delta^{\delta+1} = \Delta x^j$; explicitly

$$K^j(y^1,y^2,y^3,y^4) = \int_0^1 dt\, t^{-5}(y^j-y^1)\,[C^{(h_1)}(t^{-1}y^1, t^{-1}y^3)]^2\,\delta(y^1-y^2)\,\delta(y^3-y^4)\,.$$

The graph G_{f_2} with its subgraph G_{f_1} replaced by RG_{f_1} is then given by (the same legs are contracted as in the original graph G_{f_2})

(2.73)

These terms have $\delta = 0$ and -1 respectively, and so only the first is affected by the R-operation at f_2 (which introduces one more derivative), and we obtain

RG_{f_2} =

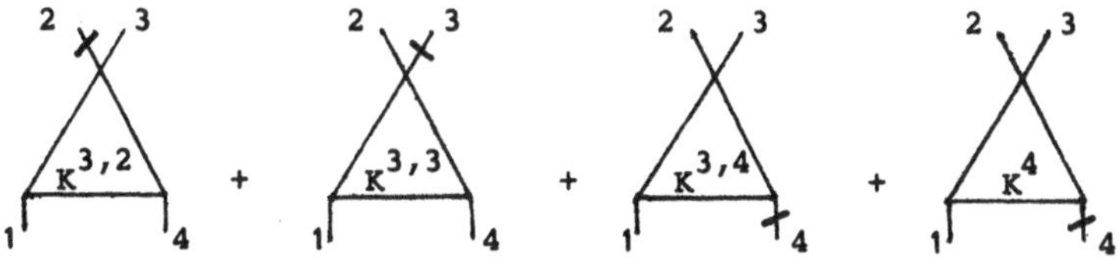

(2.74)

There are several lessons in this simple example: 1) Even if the original graph is a monomial the renormalized graph is not, because the derivatives may act on different legs. 2) Even if all terms in the expression for the renormalized subgraph have the same δ (0 in the above RG_{f_1}), this will not necessarily be the case for the output at the next fork (before renormalization) because the derivatives may or may not remain on external legs (see (2.73)). 3) All terms in the output from an R-operation have $\delta \leq -1$.

We shall adopt the convention that a graph in the renormalized tree expansion corresponds to a monomial. That is, at each R-fork f we specify a choice of one term from the sum (2.68). Similarly, at a C-fork we make a choice of one term from the sum (2.71) (i.e. we choose a specific value for the number of derivatives m and a specific action of the m derivatives on the fields). The number of such choices for the whole tree τ is bounded by $K^{|\mathcal{T}(\tau)|}$ where the maximum number of terms in (2.68) or (2.71) is $K < \infty$.

We can now specify how the contribution $V(\tau,\vec{\rho},\vec{h})$ to (2.56) is computed in terms of renormalized graphs. Let $G_o \in \mathcal{G}(\tau)$ be an unrenormalized graph associated with τ. We denote the set of renormalized graphs G corresponding to G_o by $\mathcal{G}(\tau,\vec{\rho},G_o)$ where by "corresponding to G_o" we mean that the contractions of legs to form lines and the choice of hard and soft lines in G are the same as in G_o. A single G in $\mathcal{G}(\tau,\vec{\rho},G_o)$ is determined by an inductive specification of which terms are chosen in (2.68) or (2.71). To be more explicit, we describe G_f supposing that the subgraphs G_{f_i} for the forks $f_1,\ldots,f_p$ immediately above f have been chosen. The legs of the subgraphs G_{f_i} are joined as determined by G_o to give a graph

$$\tilde{G}_f = \int K(x)\, \Pi(x)\, dx.$$

If $\delta = \delta(\tilde{G}_f) < 0$ we take

$$G_f = \begin{cases} \tilde{G}_f & \text{if } \rho_f = R \\ 0 & \text{if } \rho_f = C \,. \end{cases}$$

If $\delta \geq 0$ we choose

$$G_f = \begin{cases} \text{a monomial in (2.68) if } \rho_f = R \\ \text{a monomial in (2.71) if } \rho_f = C. \end{cases}$$

$\mathcal{G}(\tau, \vec{\rho}, G_o)$ is the set of all such possible G's. The total set of renormalized graphs is

$$\mathcal{G}(\tau,\vec{\rho}) = \bigcup_{G_o \in \mathcal{G}(\tau)} \mathcal{G}(\tau,\vec{\rho},G_o) \,.$$

Thus the set $\bigcup_{\vec{\rho}} \mathcal{G}(\tau,\vec{\rho},G_o)$ of renormalized graphs corresponding to a single unrenormalized graph G_o has at most $\tilde{K}^{|\mathcal{F}(\tau)|}$ members where the constant $\tilde{K} < \infty$.

The value $V^{\vec{h}}(G)$ of a graph $G \in \mathcal{G}(\tau,\vec{\rho})$ is determined by the rules i) - v) before (2.17) as modified by the appropriate R- and C-operations (2.68) and (2.71). In terms of graphs, the renormalized tree expansion (2.56) takes the form

$$V_r^U = V(\Phi^{(\leq r)}) + \sum_{\tau} \frac{1}{n(\tau)} \sum_{\vec{\rho}} \sum_{G \in \mathcal{G}(\tau,\vec{\rho})} \sum_{\vec{h} \in \mathcal{H}(\tau,\vec{\rho})} V^{\vec{h}}(G) \,. \tag{2.75}$$

<u>Convergence of the renormalized expansion</u>

We wish to prove that the expansion (2.75) is finite order by order in perturbation theory as $U \to \infty$. As in the unrenormalized expansion it is sufficient to show that each graph is convergent, i.e. $\sum_{\vec{h}} V^{\vec{h}}(G)$ remains bounded as $U \to \infty$. To do this, we investigate how the R- and C-operations affect the bound of Theorem 2.2.

In the Taylor operations it is the coordinate differences Δ which give the improvement in the UV power counting while the derivatives ∂_y are the price we have to pay. According to our bounds which follow, a factor Δ at a fork f with scale h introduces a coordinate difference Δx which is exponentially suppressed by a decay factor $e^{-M^h|\Delta x|}$ and hence contributes a good power counting factor M^{-h}. A derivative ∂_y, on the other hand, acts on a field which contracts at a lower fork

$f_i \leq f$ on the tree (or possibly acts on a field external to the whole graph G). By the basic bound (1.15) such action leads to a bad power counting factor M^{h_i}. These factors can be organized so that, for example, at an R-fork f to which the operation R^δ has been applied we obtain an improvement in the power counting by a factor

$$M^{-(\delta+1)(h_f-h_{\pi(f)})}.$$

As an illustration of this phenomenon, suppose that f is a maximal fork with scale $h = h_f$ so that its kernel K_f satisfies

$$|K_f(x)| \leq c_2^{\ell(G_f)} M^{h\Sigma d_\ell} e^{-M^h L_f(x)} \tag{2.76}$$

where, as in (2.22) - (2.23), Σd_ℓ is the total dimension of the contracted lines, $L_f(x)$ is the length of a tree $\mathcal{M}_f$ of hard lines that connects the x's, and c_2 is a constant independent of G (as will be all constants c_i). Then by (2.69) and (2.76)

$$\begin{aligned}\|R^\delta K_f\|_0 &\equiv \int dy\ \delta(y^o)|R^\delta K_f(y)| \\ &= \frac{1}{(\delta+1)!}\int dx\ \delta(x^o)\ |\Delta^{\delta+1}\ K_f(x)| \qquad\qquad (x = y(t^{-1})) \\ &\leq c_2^{\ell(G_f)} M^{h\Sigma d_\ell} \int dx\ \delta(x^o)|\Delta^{\delta+1}(x)|e^{-M^h L_f(x)}\end{aligned}$$

where $\Delta^{\delta+1}(x) = \prod_{i=1}^{\delta+1} x^{j_i}$ (with a sup over the suppressed indices). Here it is understood that for a δ-function in $K_f(x)$, $|\delta| = \delta$. By scaling x by M^h, we find that

$$\int dx\ \delta(x^o)\ |\Delta^{\delta+1}(x)|e^{-M^h L_f(x)} \leq c_3^{n-1}\ M^{-h(d(n-1)+\delta+1)} \tag{2.77}$$

so that

$$\|R^\delta K_f\|_0 \leq c_1^{\ell(G_f)} M^{(D(G_f)-\delta-1)h}. \tag{2.78}$$

Before stating our basic bound on a renormalized graph, we first (re)introduce some notation. The degree of divergence of a graph G is

$$D(G) = \sum_{\ell\in\mathcal{L}(G)} d_\ell - d(v(G)-1)$$

where each derivative on a line ℓ contributes 1 to d_ℓ and where graphically G has lines and vertices like those of the corresponding unrenormalized graph. We do not here adopt the "trimmed" point of view used in §6 that a C-operation at a fork f replaces G_f by a (possibly dimensionful) single vertex. For example, the ϕ_4^4 renormalized graph G associated with the tree

$\tau =$ 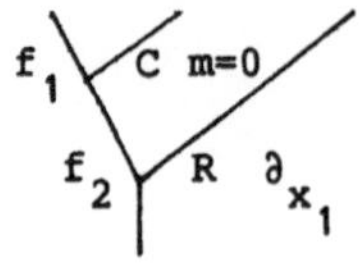

(2.79)

and corresponding to the unrenormalized graph

$G_{un} = x_1$ 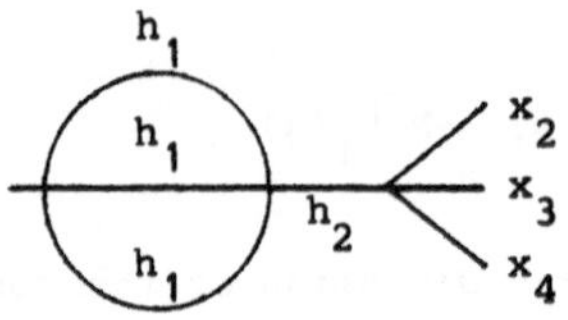

could be pictured as

$G = x_1$ 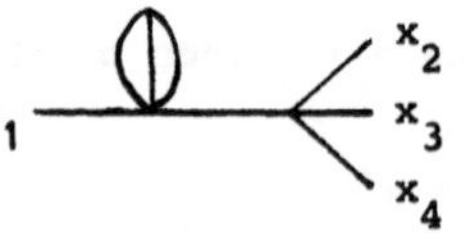

(2.80)

with 3 vertices and 4 lines like G_{un}. Thus $D(G) = 4.2 - 4(3-1) = 0$. In the trimmed point of view, G would be pictured as

$\tilde{G} =$ 

with $D(\tilde{G}) = 2 - 4(2-1) = -2$ but with the one dimensionful vertex (of dimension 2) multiplied by a coupling constant that behaves like M^{2h_2}.

We let n_f be the degree of the coordinate differences introduced in the Taylor operation at f, and we let $N_f = \sum_{f' \geq f} n_{f'}$, so that N_f is the total degree of coordinate differences in the graph G_f. For example, in the first three graphs on on the right of (2.74) $N_{f_2} = 2$ and in the last graph $N_{f_2} = 1$. As indicated above, these coordinate differences lead to good power counting factors $\approx M^{-hN_f}$. As is obvious from the definitions of the R- and C-operations, N_f also equals the total number of renormalization derivatives on the lines and legs of G_f.

If we let d_ℓ^o (or δ_λ^o) represent the dimension of a line ℓ (or leg λ) not counting renormalization derivatives, and let

$$D^o(G) = \sum_{\ell \in \mathcal{L}(G)} d_\ell^o - d(v(G)-1) \tag{2.81}$$

then clearly

$$D(G_f) - N_f = D^o(G_f) - N_f^e \tag{2.82}$$

where N_f^e is the total number of renormalization derivatives on the legs of G_f.

We now prove:

<u>Theorem 2.5</u>. Let G be a graph in the renormalized tree expansion (2.75) associated with a tree τ whose bottom fork F is labelled R or C or is unlabelled. Then the kernel of G is bounded by

$$\|G\| \leq c_0^{\ell(G)} \prod_{f \in \mathfrak{F}(\tau)} M^{(D^o(G_f)-N_f^e)(h_f-h_{\pi(f)})} \tag{2.83}$$

where c_0 is a constant independent of G and $h_{\pi(F)} = 0$.

<u>Remarks</u>. 1. The norm in (2.83) was defined in (2.21); however, in the UV regime it is simplest to take the norm to be the L_1-type norm $\|\cdot\|_0$ defined in (2.20).
2. Suppose that V is dimensionless, i.e. that every interaction vertex v in V satisfies

$$\delta(v) \equiv d - \sum_{\lambda \in \Lambda(v)} \delta_\lambda^o = 0. \tag{2.84}$$

Then

$$\begin{aligned} D^o(G_f) - N_f^e &= \sum_{\ell \in \mathcal{L}(G_f)} d_\ell^o - d(v(G_f) - 1) - N_f^e \\ &= \sum_{v \in \mathcal{V}(G_f)} \Big(\sum_{\lambda \in \Lambda(v)} \delta_\lambda^o - d\Big) - \sum_{\lambda \in \Lambda(G_f)} \delta_\lambda^o + d - N_f^e \\ &= \delta(G_f), \end{aligned} \tag{2.85}$$

and so the bound (2.83) reads

$$\|G\| \leq c_0^{\ell(G)} \prod_{f \in \mathfrak{F}(\tau)} M^{\delta(G_f)(h_f-h_{\pi(f)})}. \tag{2.86}$$

3. When it becomes necessary to distinguish between the degrees δ of a graph G_f before and after the Taylor operation at the fork f is applied we shall write G_f^u for the unlabelled graph (i.e. no R- or C-label at f) and G_f for the labelled graph. According to our renormalization prescription the number of derivatives introduced by the labelling at f is

$$n_f = \begin{cases} (\delta(G_f^u)+1)_+ & \text{if } f \in \mathfrak{F}_R \\ 0,\ 1,\ \ldots,\ \text{or } \delta(G_f^u) & \text{if } f \in \mathfrak{F}_C \end{cases}$$

where $(x)_+ = x$ if $x \geq 0$ and $(x)_+ = 0$ if $x < 0$. Therefore $\delta(G_f) = \delta(G_f^u) - n_f$ satisfies

$$\begin{aligned} \delta(G_f) &\leq -1 && \text{if } f \in \mathfrak{F}_R \\ 0 \leq \delta(G_f) &< d && \text{if } f \in \mathfrak{F}_C. \end{aligned} \tag{2.87}$$

For this reason we call (2.86) the "spring-loaded" bound. It displays the exponential decay between a fork f and its predecessor $\pi(f)$ since (recall (2.50))

$$\begin{aligned} h_f - h_{\pi(f)} &> 0 && \text{if } f \in \mathfrak{F}_R \\ h_f - h_{\pi(f)} &\leq 0 && \text{if } f \in \mathfrak{F}_C. \end{aligned}$$

Note that the exponential spring loses its stiffness in one case, namely a marginal C-fork where $\delta(G_f) = 0$. The spring-loaded bound allows us to sum over scales moving down the tree, with an accumulation of powers of h_f because of the marginal C-forks (see (2.106)).

4. For an alternate and simpler proof of this theorem using α-space, see Corollary B.4.

Proof. We establish the bound (2.83) inductively by proceeding down the tree τ and bounding the kernel K_f of each graph G_f in terms of the kernels K_{f_j} associated with the forks $f_1, \ldots, f_p$ immediately above f. (Some of these may be endpoints of τ which we regard as C-forks with $N_{f_j} = 0$.) The only complication in this inductive procedure is controlling a coordinate difference Δ as it propagates up the tree. To this end we introduce the norm

$$\|K\|_\alpha = \sup_{\tilde{\Delta}^\alpha} \int |K(y)|\ |\tilde{\Delta}^\alpha(y)|\ \ \delta(y^o)\ dy \tag{2.88}$$

where $\tilde{\Delta}^\alpha(y) = \prod_{r=1}^{\alpha} (y^{i_r} - y^{j_r})$ is any difference of degree $\alpha \geq 0$, and the sup in (2.88) takes place over all choices of indices i_r, j_r.

The inductive hypothesis on the kernel of an unlabelled fork (i.e. no Taylor operation on the fork f) is

$$\|K_f\|_\gamma \leq c_1^{\ell(G_f)}\ 2^{\gamma+d\kappa_f}\ \gamma!\ P_f\ M^{(D(G_f)-N_f-\gamma)h_f} \tag{2.89}$$

where κ_f is the number of forks $f' > f$, and

$$P_f = \prod_{f'>f} M^{(D(G_{f'})-N_{f'})(h_{f'}-h_{\pi(f')})} .$$

If a C-operation is applied to f with m additional derivatives on Π_f and a corresponding factor Δ^m in (2.71) then we have from (2.89) (with $\gamma=m$)

$$\|C_m K_f\|_0 \leq \frac{1}{m!} c_1^{\ell(G_f)} 2^{m+d\kappa_f} m! \, P_f \, M^{(D(G_f)-N_f)h_f}$$

where N_f includes the m new derivatives. Hence, since $m < d$,

$$\|C_m K_f\|_0 \leq c_1^{\ell(G_f)} 2^{d\bar{\kappa}_f} P_f \, M^{(D(G_f)-N_f)h_f} \tag{2.90}$$

where $\bar{\kappa}_f = \kappa_f+1$ is the number of forks $\geq$ f. Note that because of the strict localization of $C_m K_f$ the degree α in (2.88) must be 0.

If an R-operation is applied to f with $\delta+1$ additional derivatives on Π_f and a corresponding factor $\Delta^{\delta+1}$ in (2.69) then we find, upon making the change of variable $x = y(t^{-1})$,

$$\|R^\delta K_f\|_\alpha = \sup_{\tilde{\Delta}^\alpha} \left[\frac{1}{\delta!} \int_0^1 dt(1-t)^\delta \, t^\alpha\right] \int |\tilde{\Delta}^\alpha(x) \, \Delta^{\delta+1} K(x)| \, \delta(x^o) dx$$

$$\leq \left[\frac{\alpha!}{(\alpha+\delta+1)!}\right] c_1^{\ell(G_f)} 2^{\alpha+\delta+1+d\kappa_f} (\alpha+\delta+1)! \, P_f \, M^{(D(G_f)-N_f-\alpha)h_f}$$

by (2.89) with $\gamma = \alpha+\delta+1$ (again, N_f includes the $\delta+1$ new derivatives). Since $\delta+1 \leq d$, we obtain

$$\|R^\delta K_f\|_\alpha \leq c_1^{\ell(G_f)} 2^{\alpha+d\bar{\kappa}_f} \alpha! \, P_f \, M^{(D(G_f)-N_f-\alpha)h_f} . \tag{2.91}$$

Clearly, if we take f=F and set $c_o = c_1 2^d$ in the bounds (2.89) (with $\gamma=0$), (2.90), and (2.91) (with $\alpha=0$), then we obtain the conclusion (2.83) of the theorem.

We now assume that (2.90) and (2.91) hold for the C- and R-forks $f_1,\ldots,f_p$ immediately above f ((2.90) certainly holds for f_j an endpoint) and we deduce (2.89) for f. We write j for f_j and divide the variables y_j of $K_{f_j} = K_j$ into external variables y_j^e (those whose fields do not contract at f) and internal variables y_j^i (those whose fields do). In this notation, $y_f = y_1^e \cup \ldots \cup y_p^e$. We also write

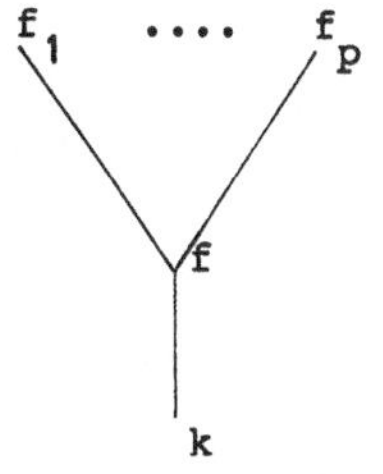

$y_f^i = y_1^i \cup \ldots \cup y_p^i$. The (unrenormalized) output from f is

$$\int \Pi dy_j \; \tilde{K}_j(y_j) \; \partial^i H_f(y_f^i) \; \Pi_f(y_f) \equiv \int dy_f \; K_f(y_f) \; \Pi_f(y_f) \tag{2.92}$$

where $\tilde{K}_j = C_{m_j} K_j$ or $R^{\delta_j} K_j$, $\Pi_f(y_f)$ is the Wick product of the uncontracted fields $\Phi^{(\leq k)}$ from $f_1, \ldots, f_p$ (including the derivatives on these fields), the kernel H_f is the product of propagators $C^{(h_f)}$ or $C^{(<h_f)}$, one for each line of g_f, such that the clusters $y_1, \ldots, y_p$ are connected by a tree $\mathcal{M}_f$ of hard lines, and ∂^i is the product of derivatives in the internal variables y_f^i.

From the definitions (2.92) and (2.88)

$$\|K_f\|_\gamma \leq \sup_{\tilde{\Delta}^\gamma} \int \prod_j dy_j \; |\tilde{K}_j(y_j)| \; |\tilde{\Delta}^\gamma(y_f)| \; |\partial^i H_f(y_f^i)| \; \delta(y_f^o). \tag{2.93}$$

We estimate the H_f factor in (2.93) by (1.15):

$$|\partial_y^i H_f(y_f^i)| \leq c_2^{\ell(g_f)} \; M^{bh_f} \; e^{-M^{h_f} L_f} \tag{2.94}$$

where L_f is the length of $\mathcal{M}_f$ and

$$b = \sum_{\ell \in \mathcal{L}(g_f)} d_\ell = D(g_f) + dp - d.$$

(d_ℓ counts derivatives on the line ℓ.)

Each difference Δx occurring in $\tilde{\Delta}^\gamma$ can be estimated by

$$|\Delta x| \leq \sum_{j=1}^{p} |\Delta y_j| + L_f \tag{2.95}$$

where each Δy_j is a difference of variables y_j^r and y_j^s in a single cluster y_j. See (B.19) for the justification of (2.95) involving appropriate t-dependence.

From (2.93) - (2.95) and the multinomial theorem we have the bound

$$\|K_f\|_\gamma \leq c_2^{\ell(g_f)} \; M^{bh_f} \; \sum_{\vec{\alpha}} \frac{\gamma!}{\vec{\alpha}!} \; \Big\| \frac{1}{\beta!} \prod_j \tilde{\Delta}^{\alpha_j} \tilde{K}_j \; L_f^\beta \; e^{-M^{h_f} L_f} \Big\|_0 \tag{2.96}$$

where the sum takes place over nonnegative integers $\alpha_1, \ldots, \alpha_p$ with $\alpha = \alpha_1 + \ldots + \alpha_p \leq \gamma$, $\beta = \gamma - \alpha$ and $\vec{\alpha}! = \Pi \alpha_j!$ (At a C-fork, $\alpha_j = 0$ because of the δ-functions in the kernel (2.71).)

We apply the inequality

$$\frac{1}{\beta!} L_f^\beta e^{-M^{h_f} L_f} \leq 2^\beta M^{-\beta h_f} e^{-M^{h_f} L_f/2}$$

and then integrate out the y_j's in (2.96) much as we did in (2.25). If $y_f^o \in y_1$, say, then the variables in y_1 are integrated last. We first integrate the y_j's which are end clusters of $\mathcal{M}_f$. If ℓ_k is the single line of $\mathcal{M}_f$ with an end-point in y_j then the integration over y_j gives

$$\int dy_j \, |\tilde{\Delta}^{\alpha_j} \tilde{K}_j| \, e^{-M^{h_f} |\ell_k|/2} \leq c_3 \, \|\tilde{K}_j\|_{\alpha_j} \, M^{-dh_f}. \tag{2.97}$$

Continuing to integrate out the new end-clusters exposed in this way, we bound the norm in (2.96) by

$$c_3^{p-1} \, 2^\beta \, M^{-(\beta+pd-d)h_f} \prod_j \|\tilde{K}_j\|_{\alpha_j}. \tag{2.98}$$

Using the fact that $h_j - h_f \geq 1$ at an R-fork f_j, we then obtain

$$\frac{2^{-\gamma}}{\gamma!} \|K_f\|_\gamma \, M^{\gamma h_f} \leq c_2^{\ell(g_f)} \, c_3^{p-1} \, M^{D(g_f)h_f} \sum_{\vec{\alpha}} M^{-\alpha} \prod_j \frac{2^{-\alpha_j}}{\alpha_j!} \|\tilde{K}_j\|_{\alpha_j} \, M^{\alpha_j h_j} \tag{2.99}$$

By (2.90) and (2.91),

$$\prod_j \frac{2^{-\alpha_j}}{\alpha_j!} \|\tilde{K}_j\|_{\alpha_j} \, M^{\alpha_j h_j} \leq \prod_j c_1^{\ell(G_j)} \, 2^{d\bar{\kappa}_j} \, P_j \, M^{(D(G_j)-N_j)h_j}$$

$$= c_1^{\ell(G_f)-\ell(g_f)} \, 2^{d\kappa_f} \, P_f \, M^{h_f \sum_j (D(G_j)-N_j)}$$

$$= c_1^{\ell(G_f)-\ell(g_f)} \, 2^{d\kappa_f} \, P_f \, M^{(D(G_j)-D(g_f)-N_f)h_f} .$$

Substituting in (2.99), we find that

$$\frac{2^{-\gamma}}{\gamma!} \|K_f\|_\gamma \, M^{\gamma h_f} \leq c_2^{\ell(g_f)} \, c_1^{\ell(G_f)-\ell(g_f)} \, 2^{d\kappa_f} \, c_3^{p-1} \, P_f M^{(D(G_f)-N_f)h_f} \sum_{\vec{\alpha}} M^{-\alpha}. \tag{2.100}$$

The sum over $\vec{\alpha}$ can be bounded by

$$\sum_{\vec{\alpha}} M^{-\alpha} \leq \left(\sum_{n=0}^{\infty} M^{-n}\right)^p \leq c_4^{\ell(g_f)} .$$

Inserting this into (2.100) gives

$$\|K_f\|_\gamma \le \left(\frac{c_2c_3c_4}{c_1}\right)^{\ell(g_f)} c_1^{\ell(G_f)} 2^{\gamma+d\kappa_f} \gamma!\, P_f\, M^{(D(G_f)-N_f-\gamma)h_f}$$

If we now choose $c_1 = c_2c_3c_4$ we obtain the desired inequality (2.89) and the theorem.

∎

For the rest of this section we assume that the interaction is dimensionless so that the spring-loaded bound (2.86) holds with $\delta(G_f)$ satisfying (2.87). We now perform the sum over scales in (2.75), starting from the top of the tree and working down. The following notational organization is often useful for this purpose (see e.g. §5):

We wish to bound the sum (uniformly in U)

$$\sum_{\vec{h}\in\mathcal{H}(\tau,\vec{\rho})} \prod_{f\in\mathcal{F}(\tau)} M^{\delta_f(h_f-h_{\pi(f)})} \tag{2.101}$$

where $\delta_f = \delta(G_f)$. Let τ_f be the subtree of τ with lowest fork f and root scale $h_{\pi(f)}$ and $\mathcal{H}_f$ the set $\mathcal{H}(\tau,\vec{\rho})$ of (2.50) restricted to the scales $\{h_{f'}|f'\ge f\}$ of τ_f. We introduce

$$\bar{U}_f(h_{\pi(f)}) = \sum_{\vec{h}\in\mathcal{H}_f} \prod_{f'\ge f} M^{\delta_{f'}(h_{f'}-h_{\pi(f')})} \tag{2.102}$$

and

$$U_f(h_f) = \sum_{\substack{\vec{h}\in\mathcal{H}_f \\ h_f \text{ fixed}}} \prod_{f'>f} M^{\delta_{f'}(h_{f'}-h_{\pi(f')})}. \tag{2.103}$$

Then

$$U_f(h_f) = \prod_{f':\pi(f')=f} \bar{U}_{f'}(h_f) \tag{2.104}$$

and

$$\bar{U}_f(h_{\pi(f)}) = \sum_{h_f} M^{\delta_f(h_f-h_{\pi(f)})} U_f(h_f)\ , \tag{2.105}$$

where the sum takes place over $h_f > h_{\pi(f)}$ if f is an R-fork and over $h_f \le h_{\pi(f)}$ if f is a C-fork.

According to (2.86), we have the following bound on the kernel of a graph G with root scale k and fork scales summed over $\mathcal{H}$:

$$\|G\| \leq c_0^{\ell(G)} \; M^{\delta(G)k} \; \bar{U}_F(k) \tag{2.106}$$

(the factor $M^{\delta(G)k}$ cancels against the factor $M^{-\delta(G_F)k}$ in the definition of $\bar{U}_F(k)$).

It is a simple matter to bound the U_f's by induction using (2.104) and (2.105): As remarked above, powers of h accumulate because of marginal C-forks. More precisely,

$$U_f(h) \leq 2^{\alpha_f} \lambda_{\beta_f}(h) \tag{2.107a}$$

and

$$\bar{U}_f(k) \leq 2^{\bar{\alpha}_f} \lambda_{\bar{\beta}_f}(k) \tag{2.107b}$$

where

$$\lambda_n(h) = \sum_{i=1}^{\infty} (h+1+i)^n \, M^{-i/2}, \tag{2.108}$$

and

$$\alpha_f = |\{f' \in \mathfrak{F}_C | \; f' > f, \; \delta_{f'} > 0\}| \qquad \bar{\alpha}_f = |\{f' \in \mathfrak{F}_C | \; f' \geq f, \; \delta_{f'} > 0\}|$$

$$\beta_f = |\{f' \in \mathfrak{F}_C | \; f' > f, \; \delta_{f'} = 0\}| \qquad \bar{\beta}_f = |\{f' \in \mathfrak{F}_C | \; f' \geq f, \; \delta_{f'} = 0\}| \; .$$

Like the power function $(h+1)^n$, $\lambda_n(h)$ satisfies (for M sufficiently large)

$$\prod_j \lambda_{n_j}(h) \leq \lambda_{\Sigma n_j}(h) \tag{2.109}$$

$$\sum_{h=k+1}^{\infty} M^{-(h-k)} \; \lambda_n(h) \leq \lambda_n(k) \tag{2.110a}$$

$$\sum_{h=0}^{k} M^{h-k} \; \lambda_n(h) \leq 2\lambda_n(k) \tag{2.110b}$$

$$\sum_{h=0}^{k} \lambda_n(h) \leq \lambda_{n+1}(k) \tag{2.110c}$$

$$\lambda_n(h) \leq n!\,(h+1)^n \; . \tag{2.111}$$

For the proof of these elementary facts, see Lemma 8.2.

Let f be a fork with forks $f_1, \ldots, f_p$ immediately above it, each with its own R or C label and values of $\delta_{f_j} = \delta_j$, $\bar{\alpha}_j$ and $\bar{\beta}_j$ (some of the f_j's may be endpoints, which we regard as C-forks with $\delta_j = \bar{\alpha}_j = \bar{\beta}_j = 0$). Assuming (2.107b) for the f_j's we deduce (2.107a) for f by using (2.104) and (2.109):

$$U_f(h) \leq \prod_j 2^{\bar{\alpha}_j} \lambda_{\bar{\beta}_j}(h) \leq 2^{\alpha_f} \lambda_{\beta_f}(h) .$$

Then, using (2.105) and (2.110) we deduce (2.107b) for f. This establishes (2.107) for all forks in τ.

From (2.106) and (2.111) we obtain, for $k \geq -1$,

$$||G|| \leq c_5^{\ell(G)} M^{\delta(G)k} \lambda_{\bar{\beta}}(k) \leq c_5^{\ell(G)} \bar{\beta}!(k+2)^{\bar{\beta}} M^{\delta(G)k} \qquad (2.112)$$

where $\bar{\beta}$ is the number of marginal C-forks in G. Similarly, for a subgraph G_f, we have

$$||G_f|| \leq c_5^{\ell(G_f)} M^{\delta(G_f)h_{\pi(f)}} \lambda_{\bar{\beta}_f}(h_{\pi(f)}) . \qquad (2.113)$$

In the "trimmed" picture, the bound (2.113) says that, up to power corrections (i.e. $\lambda_{\bar{\beta}_f}(k)$), a C-vertex at f enters the power counting with a scale-dependent coupling constant $M^{\delta k}$, where δ is the degree of the counterterm and $k = h_{\pi(f)}$. We restate (2.112) as a theorem:

<u>Theorem 2.6</u>. Consider the tree expansion (2.75) for the effective potential V_r^U for a Euclidean quantum field theory with dimensionless interaction.

a) The kernel of a renormalized graph G corresponding to a tree τ in the sum

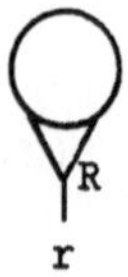

satisfies

$$||G|| \leq c_o^{\ell(G)} \kappa!(r+2)^{\kappa} M^{\delta(G)r}$$

uniformly in U, where c_o is a constant independent of G, κ is the number of marginal C-forks in τ, and $\delta(G) \leq -1$.

b) The local counterterm given by a graph G corresponding to a tree τ in the sum

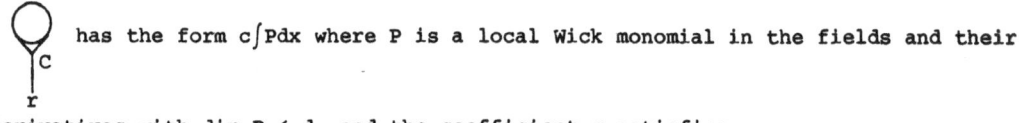

has the form $c\int P dx$ where P is a local Wick monomial in the fields and their derivatives with dim P $\le$ d, and the coefficient c satisfies

$$|c| \le c_o^{\ell(G)} \kappa!(r+1)^{\kappa} M^{(d-\dim P)r} .$$

If you allow gauge <u>variant</u> counterterms (as in (1.17)) then QED_4 is UV-renormalizable by the above theorem, and there is nothing more to be said. If you forbid gauge variant counterterms, then we invite you to read about the modification of the tree expansion in the following three sections.

§3. Loop Regularization

In this section we define the loop subtraction Λ which (together with the photon propagator cutoffs I and U) renders unrenormalized QED_4 finite, and which (unlike the electron propagator cutoff N) preserves the Ward identities. Our definition of the loop regularization is standard (see Ref. 13), but we give a careful discussion of bounds on and renormalization of loop regularized graphs.

If we formally integrate out the electron fields $\psi, \bar{\psi}$ in the basic integral defined by (1.4) and (1.12) then we obtain

$$\begin{aligned} \int e^{V(\Phi)} d\nu(\psi,\bar{\psi}) &= \det_2(1 + eS\not{A}) \\ &= \exp \operatorname{Tr}\left[\ln(1 + eS\not{A}) - eS\not{A}\right] \\ &= \exp \left[\sum_{n=2}^{\infty} \frac{(-1)^{n-1}}{n} e^n \operatorname{Tr}(S\not{A})^n \right] \end{aligned} \tag{3.1}$$

where the Wick ordering of V removes the n = 1 term. (In fact, no odd n terms occur in (3.1) by Furry's Theorem (see Lemma 3.3) and so we consider only even n.) The calculation (3.1) is not rigorous even for $A \in C_0^\infty$ because of the UV singularity of S. In momentum space we have for the n-th order "loop"

$$\operatorname{Tr}(S\not{A})^n = \int dq_1 \cdots dq_n dp\, \delta(q_1 + \ldots + q_n)\, \ell^{\vec{\mu}}(\vec{q},p,m) \prod_{i=1}^{n} \hat{A}_{\mu_i}(q_i) \tag{3.2}$$

where

$$\ell^{\vec{\mu}}(\vec{q},p,m) = (2\pi)^{-4n} \operatorname{tr} \gamma^{\mu_1} \frac{1}{\not{p}+\not{q}_1+m} \gamma^{\mu_2} \frac{1}{\not{p}+\not{q}_1+\not{q}_2+m} \cdots \gamma^{\mu_n} \frac{1}{\not{p}+\not{q}_1+\ldots+\not{q}_n+m} \tag{3.3}$$

where tr denotes the trace over spinor indices. For large p (and n even)

$$\ell^{\vec{\mu}}(\vec{q},p,m) \sim p^{-n}$$

and hence the integral over p in (3.2) diverges for n = 2 and 4.

We regularize this divergence by taking differences with respect to m. Writing $\vec{p} = (p_1,\ldots,p_n) = (p+q_1, p+q_1+q_2, \ldots, p+q_1+\ldots+q_n)$ and

$$\ell(\vec{p},m) = \ell^{\mu}(\vec{q},p,m) = \text{const.tr}\ \gamma^{\mu_1} \frac{1}{\not{p}_1+m} \gamma^{\mu_2} \frac{1}{\not{p}_2+m} \cdots \gamma^{\mu_n} \frac{1}{\not{p}_n+m}$$

we find upon taking the tr that

$$\ell(\vec{p},m) = \text{const.tr}\ \gamma^{\mu_1} \frac{-\not{p}_1+m}{p_1^2+m^2} \gamma^{\mu_2} \cdots \gamma^{\mu_n} \frac{-\not{p}_n+m}{p_n^2+m^2}$$

$$= \frac{Q_n(\vec{p}) + m^2 Q_{n-2}(\vec{p}) + \cdots}{(p_1^2+m^2)\cdots(p_n^2+m^2)} = \frac{Q_n + m^2 Q_{n-2} + \cdots}{D_{2n} + m^2 D_{2n-2} + \cdots + m^{2n}} \tag{3.4}$$

where $Q_j(\vec{p})$ and $D_j(\vec{p})$ are polynomials in $\vec{p}$ of degree j that are independent of m. (The calculation (3.4) exploits the fact that the trace of a product of an odd number of γ's vanishes.) The difference

$$\delta\ell(\vec{p},m,\tilde{m}) \equiv \ell(\vec{p},m) - \ell(\vec{p},\tilde{m})$$

$$= \frac{(Q_n+m^2Q_{n-2}+\cdots)(D_{2n}+\tilde{m}^2D_{2n-2}+\cdots)-(Q_n+\tilde{m}^2Q_{n-2}+\cdots)(D_{2n}+m^2D_{2n-2}+\cdots)}{D_{2n}^2+\cdots}$$

$$= \frac{(\tilde{m}^2-m^2)(Q_nD_{2n-2}-Q_{n-2}D_{2n}) + \cdots}{D_{2n}^2+\cdots} \tag{3.5}$$

is then $O\left(\frac{1}{p^{n+2}}\right)$ for large p.

We can repeat this procedure to reduce the degree in p by 2 again by setting

$$\delta^2\ell(\vec{p},m,\tilde{m},m',\tilde{m}') = \delta\ell(\vec{p},m,\tilde{m}) - \delta\ell(\vec{p},m',\tilde{m}')$$

provided

$$\tilde{m}^2 - m^2 = (\tilde{m}')^2 - (m')^2 \tag{3.6}$$

so that the coefficients of the leading term in the numerators match (see (3.5)). The simplest choice to make is $\tilde{m}^2 = (m')^2$ and to write

$$(\tilde{m}')^2 = m^2 + 2\Lambda^2 \equiv M_1(\Lambda)^2$$

$$(\tilde{m})^2 = m^2 + \Lambda^2 \equiv M_2(\Lambda)^2$$

$$m' = M_3(\Lambda) = M_2(\Lambda)$$

$$m = M_0.$$

In this notation the mass condition (3.6) takes the form

$$M_0^2 + M_1^2 = M_2^2 + M_3^2. \tag{3.7}$$

Then the second difference

$$\delta_\Lambda^2 \ell(\vec{p},m) \equiv \ell(\vec{p},m) - 2\ell(\vec{p},M_2) + \ell(\vec{p},M_1) \tag{3.8}$$

is $O(\frac{1}{p^{n+4}})$, and the integral over p in (3.2) will converge if $\vec{\ell}^\mu$ is replaced by the regularized expression $\delta_\Lambda^2 \ell$.

All the loops in (3.1) will thus be regularized by replacing $\det_2(m) = \det_2\big(1+eS(m)\not{A}\big)$ by

$$\det_\Lambda(1+eS\not{A}) \equiv \frac{\det_2(m)\det_2(M_1)}{\det_2(M_2)^2}$$

$$= \exp\Big[\sum_{n=2}^{\infty} \frac{(-1)^{n-1}}{n} e^n \delta_\Lambda^2 \operatorname{Tr}\big(S(m)\not{A}\big)^n\Big]. \tag{3.9}$$

A convenient way to implement this regularization is to introduce fictitious fields ψ_i, $\bar{\psi}_i$, i=1,2,3, where ψ_1, $\bar{\psi}_1$ are fermi fields and ψ_2, $\bar{\psi}_2$, ψ_3, $\bar{\psi}_3$ are spinor <u>bose</u> fields with propagators

$$\langle\psi_i\bar{\psi}_i\rangle = S(M_i). \tag{3.10}$$

Expectations are defined using the Lagrangian

$$\mathcal{L}_\Lambda = \mathcal{L} + \sum_{i=1}^{3} :\bar{\psi}_i(-i\not{\partial}+M_i+e\not{A})\psi_i:. \tag{3.11}$$

In terms of the formalism of §1 the fields are

$$\Phi = (\Phi_1,\ldots,\Phi_9) \equiv (A,\psi,\bar{\psi},\psi_1,\ldots,\bar{\psi}_3) \equiv (A,\Psi); \tag{3.12a}$$

the covariance C_Λ has non-zero elements (T denotes transpose)

$$(C_\Lambda)_{11} = D \qquad (C_\Lambda)_{23} = -(C_\Lambda)^T_{32} = S(m)$$

$$(C_\Lambda)_{45} = -(C_\Lambda)^T_{54} = S(M_1)$$

$$(C_\Lambda)_{67} = (C_\Lambda)_{89} = (C_\Lambda)^T_{76} = (C_\Lambda)^T_{98} = S(M_2);$$

the (formal) free measure is given by

$$dP_\Lambda(\Phi) = \text{const.}\ e^{-\frac{1}{2}\Phi C_\Lambda^{-1}\Phi} \prod_i \mathfrak{d}\Phi_i \tag{3.12b}$$

$$\equiv \text{const.}\ d\mu(A)\ d\nu_\Lambda(\Psi);$$

and the interaction potential is

$$V_\Lambda(\Phi) = -e \sum_{i=0}^{3} \int :\bar\psi_i \not{A}\, \psi_i:dx \equiv -e\int :\Psi\not{A}\Psi:dx \tag{3.12c}$$

where $\psi_0 = \psi$, $\bar\psi_0 = \bar\psi$.

The effect of the bosonic nature of $\psi_2, \ldots, \bar\psi_3$ is to give a sign change for each loop, or in terms of the determinant calculation of (3.1) and (3.9):

$$\int e^{V_\Lambda(\Phi)} d\nu_\Lambda(\Psi) = \frac{\det_2(M_0)\det_2(M_1)}{\det_2(M_2)\det_2(M_3)} = \det_\Lambda(1+eS\not{A}). \tag{3.13}$$

In general, in the version of QED defined by (3.12) an unrenormalized Feynman graph G consists of spinor lines or loops of type $i = 0, 1, 2$ or 3 connected by photon lines. Figure 3.1 shows a typical example with two spinor loops and one spinor line. By a <u>loop regularized graph</u> G_Λ we mean the sum of graphs with the same structure as G, summed over types $i = 0, 1, 2, 3$ for each loop of G (no sum over types for the spinor <u>lines</u> in G).

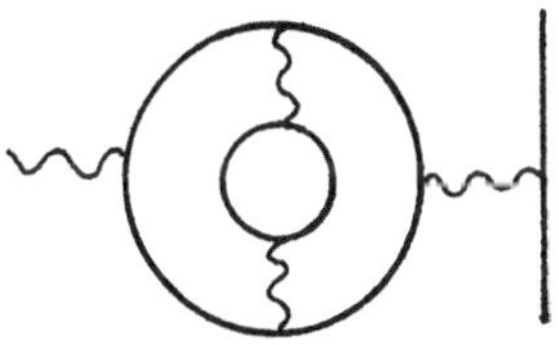

Fig. 3.1

Consider now a loop regularized graph G_Λ arising from the computation of the unrenormalized effective potential in the model (3.12):

$$V^{U,\Lambda,N}_{k,un}\left(\Phi^{(\leq k)}\right) = \log \int e^{V_\Lambda\left(A^{(\leq U)},\Psi^{(\leq N)}\right)} dP_\Lambda\left(A^{(k,U]},\Psi^{(k,N]}\right) + \text{const.} \tag{3.14}$$

Recall that until §6 an IR cutoff I=0 is in effect and that $k \geq 0$. Note also that we have imposed an UV cutoff N on all the spinor fields Ψ which is independent of the UV cutoff U on A. As indicated by the momentum space considerations in (3.2)-(3.8) the cutoff N is unnecessary:

Lemma 3.1. For U, $\Lambda < \infty$ fixed, G_Λ is convergent, uniformly in N.

Proof. We apply the unrenormalized power counting (2.35)-(2.36) to bound a graph G but we bound the contribution of each photon propagator by

$$|D^{(h)}(x,y)| \leq \text{const.}\ M^{2U} e^{-M^h|x-y|}.$$

The degree of divergence of a subgraph G_f with n vertices, ℓ_s spinor lines and λ_s spinor legs then contains no contributions from photon lines (compare with (2.36):

$$\begin{aligned} D_s(G_f) &= 3\,\ell_s - 4(n-1) \\ &= 3\,(n-\lambda_s/2) - 4n + 4 \\ &= 4 - \frac{3}{2}\lambda_s - n. \end{aligned}$$

Since λ_s is even and $n \geq 2$, $D_s(G_f) \geq 0$ iff $\lambda_s = 0$ and n = 2 or 4. The corresponding G_f's,

(3.15)

are the only subgraphs that potentially cause divergence in the bound (2.35).

As in (3.13) these subgraphs occur in (3.14) in loop regularized form $G_{f,\Lambda}$ and so to prove the lemma it is sufficient to show that for n = 2 and 4 the loop expression

$$L(\vec{x},\vec{h},m) = \text{tr} \prod_{i=1}^{n} S^{(h_i)}(x_i,x_{i+1};m)\gamma^{\mu_i},$$

where $x_{n+1} = x_1$ satisfies

$$|\delta_\Lambda^2 L| \leq \text{const. } M^{-4h} B(\vec{x},\vec{h}) \tag{3.16}$$

where $h = \min(h_1,\ldots,h_n)$ and B is the usual bound on the propagators (see (1.15a)):

$$B(\vec{x},\vec{h}) = \prod_{i=1}^{n} M^{3h_i} e^{-M^{h_i}|x_i - x_{i+1}|}.$$

In other words, the second difference δ_Λ^2 gives an improvement of M^{-4h} in the bound (the const. is of course Λ dependent). If one of the lines in the loop is soft, h can be smaller than the fork scale h_f. Nonetheless, we can replace the factor M^{3h_i} from one line by M^0 and thus $D_s(G_{f,\Lambda}) = D_s(G_f) - 3 < 0$, as desired for convergence.

To establish (3.16) for any (even) n (and any power δ_Λ^r) define $m(s) = (m^2+s\Lambda^2)^{1/2}$ so that the first difference of a function $f(m^2)$ is

$$\delta_\Lambda f = f(m^2(1)) - f(m^2(0)) = \int_0^1 ds\, f'\big(m^2(s)\big)\Lambda^2$$

and the second difference is

$$\delta_\Lambda^2 f = f(m^2(2)) - 2f(m^2(1)) + f(m^2(0))$$

$$= \Lambda^4 \int_0^1 ds \int_0^1 dt\, f''(m^2(s+t)). \tag{3.17}$$

By (3.17), (3.16) reduces to showing that

$$\left|\left(\frac{\partial}{\partial m^2}\right)^2 L\right| \leq \text{const. } M^{-4h} B(\vec{x},\vec{h}). \tag{3.18}$$

The proof of (3.18) is based on the observation that each derivative $\frac{\partial}{\partial m^2}$ produces a factor M^{-2h_i} when it acts on the ith propagator $S^{(h_i)}$. From the definition (1.13) of $S^{(h)}$,

$$L = (2\pi)^{-4n}\int\ldots\int dk_1\ldots dk_n\left[\text{tr } \Pi(-\not{k}_j+m)\,\gamma^{\mu_j}\right]\left[\Pi\zeta^{h_j}(k_j,m)\right]\,\Pi e^{ik_j(x_j-x_{j+1})}$$

$$\equiv (2\pi)^{-4n}\int dk\; T\, Z\, \Pi\, e^{ik_j(x_j-x_{j+1})}, \tag{3.19}$$

defining T and Z, where (for h > 0)

$$\zeta^h(k,m) = \int_{M^{-2h}}^{M^{-2h+2}} d\alpha \; e^{-\alpha(k^2+m^2)} .$$

Now

$$\left(\frac{\partial}{\partial m^2}\right)^2 TZ = \left[\left(\frac{\partial}{\partial m^2}\right)^2 T\right]Z + 2\left(\frac{\partial}{\partial m^2}T\right)\left(\frac{\partial}{\partial m^2}Z\right) + T\left(\frac{\partial}{\partial m^2}\right)^2 Z.$$

Since

$$\left(\frac{\partial}{\partial m^2}\right)^r \zeta^h = \left(-M^{-2h}\right)^r \int_{M^{-2h}}^{M^{-2h+2}} d\alpha \; (\alpha M^{2h})^r \; e^{-\alpha(k^2+m^2)}$$

$$\equiv M^{-2hr} \zeta_r^h \tag{3.20}$$

we have

$$\frac{\partial}{\partial m^2} Z = \sum_j M^{-2h_j} \zeta_1^{h_j} \prod_{i \neq j} \zeta^{h_i} ,$$

and similarly for $\left(\frac{\partial}{\partial m^2}\right)^2 Z$. Each $\frac{\partial}{\partial m^2}$ acting on Z explicitly supplies a factor of M^{-2h}. As for T, since the trace of the product of an odd number of γ's vanishes,

$$T = \mathrm{tr} \prod (-\not{k}_j + m) \gamma^{\mu_j}$$

$$= \left[Q_n(\vec{k}) + m^2 Q_{n-2}(\vec{k}) + \ldots + m^n Q_0\right] \tag{3.21}$$

where Q_j is a polynomial of degree j in $\vec{k} = (k_1,\ldots,k_n)$ which is at most linear in each k_j. Hence $\left(\frac{\partial}{\partial m^2}\right)^r T$ is a polynomial of degree n-2r in $\vec{k}$, at most linear in each k_i.

When we insert these expressions into the integral for $\left(\frac{\partial}{\partial m^2}\right)^2 L$ we obtain a sum of integrals each of which factors as

$$\text{const.} \prod_j \int dk_j (k_j^{\nu_j})^{a_j} M^{-2r_j h_j} \zeta_{r_j}^{h_j} e^{ik_j(x_j - x_{j+1})} \tag{3.22}$$

where $a_j = 0$ or 1 and

$$\Sigma(a_j - 2r_j) \leq n - 4. \tag{3.23}$$

By (3.20) each of the k_j-intervals in (3.22) has the form (we drop the subscripts j and write x for $x_j - x_{j+1}$):

$$\int dk(k^\nu)^a \int_{M^{-2h}}^{M^{-2h+2}} d\alpha\ (\alpha M^{2h})^r\ e^{-\alpha(k^2+m^2)} + ikx$$

$$= (2\pi)^2(-i\frac{\partial}{\partial x^\nu})^a \int_{M^{-2h}}^{M^{-2h+2}} d\alpha\ (\alpha M^{2h})^r\ \alpha^{-2}\ e^{-\alpha m^2 - x^2/4\alpha} \quad . \tag{3.24}$$

Now $\alpha M^{2h} \leq M^2$ and, for positive constants $c_i(M)$,

$$|(-i\frac{\partial}{\partial x^\nu})^a\ e^{-x^2/4\alpha}| \leq c_1\alpha^{-a/2}\ e^{-c_2M^{2h}x^2}$$

$$\leq c_3M^{ah}\ e^{-M^h|x|} \ .$$

Hence

$$|(3.24)| \leq c_5\ M^{(2+a)h}\ e^{-M^h|x|} \tag{3.25}$$

and consequently

$$|(3.22)| \leq c_6 \prod_j M^{(2+a_j-2r_j)h_j}\ e^{-M^{h_j}|x_j-x_{j+1}|}$$

$$\leq c_6 \prod_j M^{(a_j-2r_j-1)h_j}\ B(\vec{x},\vec{h}) \ .$$

Now since $a_j-2r_j-1 \leq 0$, we have

$$\Sigma\ (a_j-2r_j-1)h_j \leq h\ \Sigma(a_j-2r_j-1) \leq -4h$$

by (3.23). This completes the proof of (3.18) and the lemma. ∎

Thus, for $U,\Lambda < \infty$ fixed, the unrenormalized effective potential is finite in perturbation theory. In order to take $U,\Lambda \to \infty$ we must first renormalize. We do so now.

There is a drawback to loop regularization that is perhaps not widely recognized, namely, its fragility with respect to renormalization. We illustrate the difficulty with a simple example expressed in conventional momentum space language.

The loop of Fig. 3.2 has a kernel

$$K(q_1) = \text{const.} \int \frac{dq}{q^2} \int dp \; \mathrm{tr} \frac{1}{\not p+m} \gamma^{\mu_1} \frac{1}{\not p+\not q_1+m} \gamma^{\mu_2} \frac{1}{\not p+m} \gamma^{\nu} \frac{1}{\not p-\not q+m} \gamma^{\nu}.$$

(ν is summed over.)

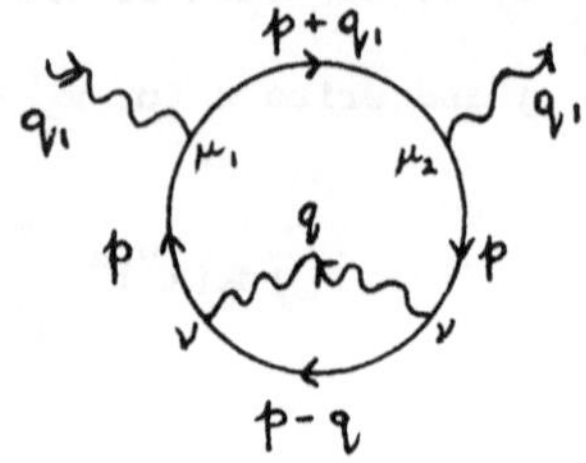

Fig. 3.2

Loop regularization and an UV cutoff on the photon propagator ($|q| \leq U$) render K finite. Now the subdiagram

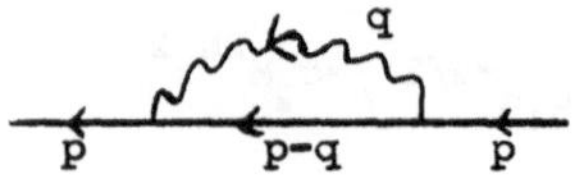

is renormalized by a counterterm

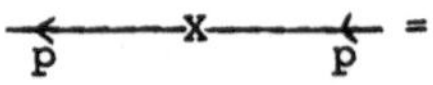

$$= \text{const.} \; (a\not p+bm) \tag{3.26}$$

where the constants $a(U,m)$ and $b(U,m)$ can be explicitly determined and are finite for $U < \infty$. The renormalization of the graph of Fig. 3.2 thus involves the graph of Fig. 3.3 which has kernel

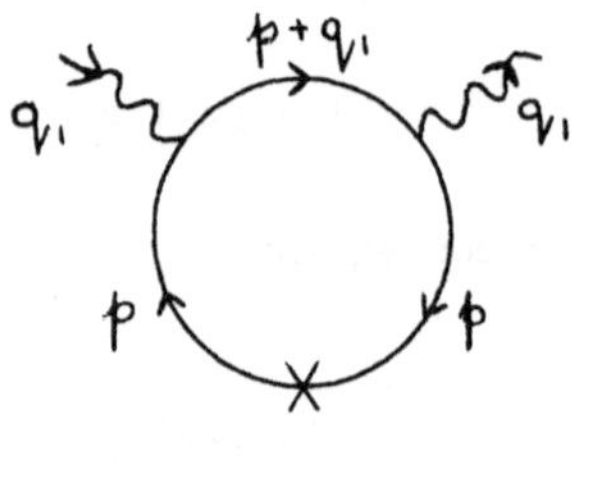

Fig. 3.3

$$L(q_1) = \text{const.} \int dp \; \mathrm{tr} \; \frac{1}{\not p+m} \gamma^{\mu_1} \frac{1}{\not p+\not q_1+m} \gamma^{\mu_2} \frac{1}{\not p+m} (a\not p+bm).$$

<u>Loop regularization fails to render L finite</u>. The reason is that a and b depend on m; hence, differences with respect to m (or differentiations $\frac{\partial}{\partial m^2}$, as in the proof of Lemma 3.1) can act on a and b in which case they do us no good in reducing the degree of the integrand in p. On the other hand, if a and b did not depend on m then the argument in (3.4)-(3.8) or in Lemma 3.1 would work and $\delta_\Lambda^2 L$ would be finite.

We are thus obliged to choose "wrong" counterterms for the graphs involving external fictitious fields, i.e., of type

$$\bar{\psi}_i\psi_i,\ \bar{\psi}_i\partial\!\!\!/\psi_i,\ \bar{\psi}_i A\!\!\!/\psi_i,\ i = 1,\ 2,\ 3. \tag{3.27}$$

For example, if the spinor lines in (3.26) are of type i we modify the value of the counterterm to be

$$\underset{p}{\text{———}}\!\!\times\!\!\underset{p}{\text{———}} = \text{const.}\ \left(a(U,M_0)p\!\!\!/ + b(U,M_0)M_i\right) \tag{3.28}$$

where $M_0 = m$ is the bare electron mass. In general, we see from (3.9) and (3.16) that the loop regularization will regularize provided that the counterterms up to any order n have the form

$$\delta V_{\Lambda,\leq n} = -\int \sum_{j=0}^{3} :\bar{\psi}_j^{(\leq N)} \left(a_n(-i\partial\!\!\!/) + b_n M_j + c_n A\!\!\!/^{(\leq U)}\right) \psi_j^{(\leq N)} : dx$$

$$+ W_n\left(A^{(\leq U)}\right) \tag{3.29}$$

where the coefficients a_n, b_n, c_n are <u>independent of j</u>, and W_n consists of the counterterms not involving external spinor fields. a_n, b_n, c_n and W_n are Λ-dependent because of internal spinor loops. We shall write (3.31) as in (3.12c):

$$\delta V_{\Lambda,\leq n} = -\int :\Psi \left(a_n(-i\partial\!\!\!/) + b_n \mathbf{M} + c_n A\!\!\!/\right) \Psi : dx + W_n(A) \tag{3.30}$$

where $\mathbf{M}$ is the 4×4 diagonal matrix with $M_{ij} = \delta_{ij}M_j$ and where we drop superscripts if there is no confusion. Note also that by Euclidean covariance the coefficients a_n, b_n, c_n are constants, i.e. they do not depend on spinor indices, and that W_n consists of the last four terms in (2.37).

The counterterms are chosen so that the diagrams that do not have external fictitious fermion lines <u>are</u> renormalized correctly. Hence, for example, $(b_{n+1}-b_n)M_0$ is the sum of all diagrams having external fields $\bar{\psi}_0\ \psi_0$, having order e^{n+1}, built with the usual interaction vertex and counterterm vertices from (3.29), and evaluated at zero external momentum.

This choice of counterterms may be implemented through a modified localization operator L^Λ. For graphs G with no fictitious field legs or with $LG = 0$, we take $L^\Lambda G = LG$. For a graph G_i with external legs as in (3.27) let G be the graph which is identical to G_i except that the spinor line connecting the external spinor legs is replaced by a real fermion line. Then we define $L^\Lambda G_i$ in terms of LG (as defined in (2.47)):

$$(L^\Lambda G_i)(A,\psi_i,\bar{\psi}_i) = \sum_{\delta \geq 0} \left(\frac{M_i(\Lambda)}{M_0}\right)^\delta (L_\delta G)(A,\psi_i,\bar{\psi}_i) \ , \tag{3.31}$$

where $L_\delta G$ contains the monomials of degree δ. The factor $\left(\frac{M_i(\Lambda)}{M_0}\right)^\delta$ is required to produce the form (3.29). See, for example, (3.28). Now the renormalized tree expansion which preserves loop regularization is of the standard form with R and C defined using the localization L^Λ in place of L.

Consider, for example, the graph G

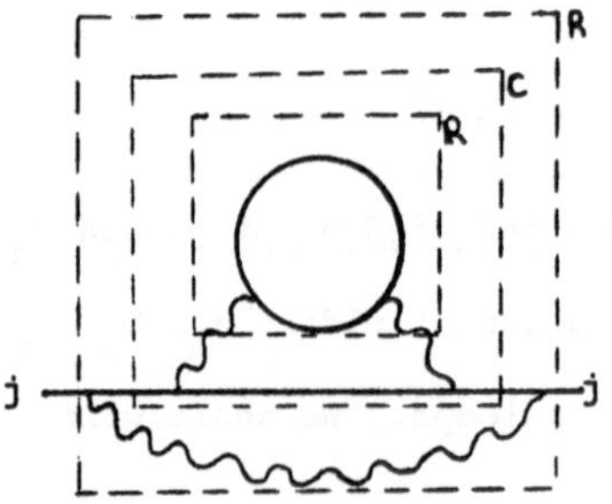

containing the boxed subgraphs $G_1 \subset G_2$. The R operation on G_1 subtracts the usual counterterms with 2 photon legs (but the regularized loop and hence these counterterms are Λ-dependent). The C operation on G_2 produces counterterms of the form (3.28)

$$\int :\bar{\psi}_j\left(a(-i\not{\partial}) + bM_j\right) \psi_j: \, dx \tag{3.32}$$

where a and b do <u>not</u> depend on j (but do depend on Λ because G_1 does). Finally, the counterterms involved in the R operation on G are computed with j set equal to 0 in the contribution (3.32) from G_2, and so, by (3.31), they again have the form (3.32) with a and b depending on Λ but not on j.

At first thought it seems disastrous to choose wrong counterterms for the fictitious fields. However, the role of these counterterms is not to effect a renormalization cancellation but rather to maintain the loop regularization. The fictitious fields and their wrong counterterms will be removed by the limit $\Lambda \to \infty$ (§5) before the cancellations between graphs and spinor counterterms need to be considered.

It is now easy to establish that the renormalization convention (3.31) maintains loop regularization:

<u>Theorem 3.2</u>. Let G_Λ be a loop regularized graph arising in the renormalized tree expansion for $V_k^{U,\Lambda,N}$. For $U, \Lambda < \infty$, G_Λ is convergent, uniformly in N. Consequently, $V_k^{U,\Lambda} = \lim_{N\to\infty} V_k^{U,\Lambda,N}$ and the counterterms $\delta V^{U,\Lambda} = \lim_{N\to\infty} \delta V^{U,\Lambda,N}$ are finite in perturbation theory.

<u>Proof</u>. Suppose that we have proved the theorem for graphs of $O(e^n)$. Let G_Λ be a graph contributing to $V_k^{U,\Lambda,N}$ which is of order n+1 in e and which has been renormalized to $O(e^n)$ but not to $O(e^{n+1})$. Then G_Λ occurs in

$$V_{k,n}^{U,\Lambda,N} \equiv \log \int e^{V_\Lambda + \delta V_{\Lambda,\leq n}} d\nu_\Lambda(\Psi^{(k,N]})\, d\mu(A^{(k,U]}) + \text{const.} \tag{3.33}$$

where $\delta V_{\Lambda,\leq n}$ consists of all counterterms of $O(e^n)$ and by our counterterm convention has the form (3.31).

By the inductive hypothesis $\delta V_{\Lambda,\leq n}$ is finite uniformly in N since it is obtained from graphs of $O(e^n)$. In the evaluation of (3.33) we treat $\delta V_{\Lambda,\leq n}$ as a contribution to the interaction. The terms $-c_n \int :\bar\Psi \not{A} \Psi:$ and $W_n(A)$ in $\delta V_{\Lambda,\leq n}$ cause no new problems; it remains to consider the effect on the power counting of the term

$$Q = -\int :\bar\Psi\big(a_n(-i\not{\partial}) + b_n M\big)\Psi:\, dx \ .$$

The insertion of Q on a spinor line of type j is harmless since instead of a propagator $S_j^{(h)}(x_1,x_2)$ we obtain

$$-\int S_j^{(h)}(x_1,x)\big(a_n(-i\not{\partial}) + b_n M_j\big)\, S_j^{(h')}(x,x_2)\, dx \tag{3.34}$$

which by (1.15a) is bounded by $\big(h_0 = \min(h,h')\big)$

$$c_1 \int M^{3h} e^{-M^h|x_1-x|} (a_n M^{h_0} + b_n M_j) M^{3h'} e^{-M^{h'}|x-x_2|}\, dx$$

$$\leq c_2 M^{3h_0} e^{-c_3 M^{h_0}|x_1-x_2|}$$

where $c_3<1$. Thus (3.34) gives essentially the same contribution to the power counting as a propagator $S_j^{(h_0)}(x_1,x_2)$.

Suppose the insertion of Q occurs in a spinor loop (see (3.19)), say, between x_1 and x_2. Then the factor

$$(-\not{k}_1+m)\gamma^{\mu_1}\zeta^{h_1}(k_1,m)$$

in (3.19) is replaced by

$$-(-\not{k}_1+m)\gamma^{\mu_1}\zeta^{h_1}(k_1,m)(a_n\not{k}_1+b_n m)(-\not{k}_1+m)\zeta^{h_1'}(k_1,m) \quad . \tag{3.35}$$

We have to modify the proof that loop regularization produces an improvement of M^{-4h} (see (3.18)); but the modification is not serious and briefly goes as follows:

Due to (3.35) the trace (3.21) over spinor indices becomes

$$T = Q_{n+2}(\vec{k}) + m^2 Q_n(\vec{k}) + \ldots + m^{n+2} Q_0$$

where Q_j is at most cubic in k_1 and at most linear in $k_2, \ldots, k_n$. The integrals over $k_2, \ldots, k_n$ in (3.22) are unchanged and are estimated as before. The integral over k_1 becomes

$$\int dk_1 P_{a_1}(k_1) \int_{M^{-2h_1}}^{M^{-2h_1+2}} d\alpha_1 \int_{M^{-2h_1'}}^{M^{-2h_1'+2}} d\alpha_1'(\alpha_1 M^{2h_1})^{r_1}(\alpha_1' M^{2h_1'})^{r_1'} e^{-(\alpha_1+\alpha_1')(k_1^2+m^2)+ik_1x} \tag{3.36}$$

where P_{a_1} is a polynomial of degree $a_1 \le 3$, $x = x_1 - x_2$, and as in (3.23)

$$\Sigma(a_j - 2r_j) - 2r_1' \le n-2. \tag{3.37}$$

We estimate (3.36) much as we did (3.24) to obtain

$$|(3.36)| \le c_1 M^{a_1 h_0} e^{-c_2 M^{h_0}|x|} \tag{3.38}$$

where $h_0 = \min(h_1, h_1')$. The gain of (3.38) over (3.25) exactly compensates for the loss of (3.37) compared to (3.23), and we deduce the desired bound (3.18).

Thus loop regularization is maintained and G_Λ is bounded uniformly in N. Hence so are the counterterms of $O(e^{n+1})$. This verifies the convergence of graphs of $O(e^{n+1})$. ∎

We conclude this section with a proof of Furry's Theorem in the path integral formalism. Let Γ be a 4x4 "charge conjugation" matrix which satisfies

$$\Gamma\gamma_\mu\Gamma^{-1} = -\gamma_\mu^t \tag{3.39}$$

where $\gamma_{ij}^t = \gamma_{ji}$. (We use the notation S^T for kernels for which there is also a transposition of spatial coordinates, i.e. $S_{ij}^T(x,y) = S_{ji}(y,x)$.) It follows from (3.39) that

$$\Gamma S\Gamma^{-1} = \frac{1}{(2\pi)^4}\int (-\not k^t+m)^{-1}\, e^{ik(x-y)}dk = S^T. \tag{3.40}$$

The following argument applies for virtually any choice of regularization of S, so long as (3.40) is preserved. Since we wish to use the regularization (1.13) and loop regularization, we give the argument for the effective potential (3.14) with these regularizations.

Corresponding to the basic transformation

$$A' = -A, \qquad \psi' = -(\Gamma^t)^{-1}\bar\psi, \qquad \bar\psi' = \Gamma\psi$$

define the change of variables

$$\Phi_i' = \begin{cases} -\Phi_1 & i = 1 \\ -(\Gamma^t)^{-1}\Phi_{i+1} & i = 2, 4, 6, 8 \\ \Gamma\Phi_{i-1} & i = 3, 5, 7, 9. \end{cases} \tag{3.41}$$

Then it is easy to verify that the covariance is unchanged:

$$C_{ij}' = \langle\Phi_i'\Phi_j'\rangle = C_{ij}. \tag{3.42}$$

For example,

$$C_{23}'(x,y) = -(\Gamma^t)^{-1}\langle\Phi_3(x)\Phi_2(y)\rangle\,\Gamma^t \qquad \text{(by (3.42))}$$

$$= (\Gamma^t)^{-1}S^t(y,x)\Gamma^t \qquad \text{(anticommutativity)}$$

$$= S(x,y) = C_{23}(x,y). \qquad \text{(by (3.40))}$$

The basic interaction is also invariant since

$$\Psi_{2i+1}^t\not A\Psi_{2i} = (\Psi_{2i}'^t\,\Gamma)\not A'\,\Gamma^{-1}\Psi_{2i+1}' \qquad \text{(by (3.41))}$$

$$= -\Psi'^{t}_{2i} \not{A}'^{t} \Psi'_{2i+1} \qquad \text{(by (3.39))}$$

$$= \Psi'^{t}_{2i+1} \not{A}' \Psi'_{2i} \qquad \text{(anticommutativity)}.$$

From the change of variables (3.41) in the integral (3.14) we conclude that the unrenormalized effective potential is invariant:

$$V^{U,\Lambda,N}_{k,un}\left(\Phi'^{(\leq k)}\right) = V^{U,\Lambda,N}_{k,un}\left(\Phi^{(\leq k)}\right). \tag{3.43}$$

By (2.60) the counterterms and the renormalized effective potential are also invariant:

<u>Lemma 3.3</u>. (Furry's Theorem[25]) The covariance, the interaction potential, the counterterms, and the unrenormalized and renormalized effective potentials are all invariant under the transformation (3.41).

In particular, the photon counterterms $W_n\left(A^{(\leq U)}\right)$ and the loops $\mathrm{Tr}(S\not{A})^n$ of (3.2) must be even in A.

§4. Ward Identities

We now prove that for renormalized QED, with photon cutoffs U and I, and loop regularization Λ, Ward identities hold and the counterterms δV are gauge invariant.

From Theorem 3.2, the order e^n counterterms δV_n (taken with spinor cutoff $N = \infty$) are finite and have the form

$$\delta V_n(\Phi) = -\int :\bar{\Psi}\left(a_n(-i\partial\!\!\!/)+b_n M+c_n A\!\!\!/\right)\Psi:dx + \delta W_n(A) \tag{4.1}$$

for (real) numbers a_n, b_n, c_n which are independent of the fictitious field index i. Here, and in the following, we drop the fixed and finite labels I, U, Λ, and write $\Phi \equiv (A,\bar{\Psi},\Psi)$ for $(A,\bar{\psi}_i, \psi_i)_{i=0,1,2,3}$. We also write **M** for the spinor mass matrix, $S \equiv (-i\partial\!\!\!/1+M)^{-1}$ for the spinor covariance, and let $\delta V_{\leq n} \equiv \sum_{k=2}^{n} \delta V_k$ denote the counterterms up to order e^n, $a_{\leq n} = \sum_{k=2}^{n} a_k$ etc.

<u>Theorem 4.1:</u> For each $k\geq 2$, δV_k has gauge invariant form. That is, δV_k is of the form (4.1) with

$$\delta W_k(A) = -d_k\int :F^2:dx + \text{const.} \tag{4.2a}$$

for some real number d_k, and in addition satisfies the "$Z_1=Z_2$ condition":

$$c_k = ea_{k-1},\ c_2 = 0. \tag{4.2b}$$

Condition (4.2a) is the condition for a gauge invariant renormalization, pure and simple. Condition (4.2b) goes beyond the minimal requirement of gauge invariance and says that the electric charge used in the definition of gauge transformations is unchanged order by order in perturbation theory.

We prove Theorem 4.1 in the case where I = 0. In §7 we outline the (straightforward) extension to the case when I < 0.

We will use a Ward Identity expressed in terms of the <u>effective potential renormalized up to order e^n</u>:

$$V_{e,n}(\Phi^e) \equiv \log \int e^{(V+\delta V_{\le n})(\Phi+\Phi^e)} dP(\Phi) + \text{const.} \tag{4.3}$$

The counterterms of order e^{n+1} are given by

$$\delta V_{n+1} = -L^{\Lambda}_{n+1} V_{e,n}. \tag{4.4}$$

as in (3.31). We prove Theorem 4.1 by the following inductive strategy: Assuming $\delta V_{\le n}$ is of gauge invariant form, we establish a Ward Identity for $V_{e,n}$ (Lemma 4.2). This leads to a local Ward Identity for δV_{n+1} (Lemma 4.3) from which we conclude that δV_{n+1} has gauge invariant form (Theorem 4.1).

__Lemma 4.2.__ Suppose V and $\delta V_{\le n}$ have gauge invariant form. Then for any smooth polynomially bounded function $\chi(x)$,

$$V_{e,n}(\Phi^e) = V_{e,n}\left(A^e + \alpha_{n-1}^{-1}\alpha_n \partial\chi,\ (1+\mathbf{SX})e^{-ie\chi}\Psi^e,\ \bar{\Psi}^e e^{ie\chi}(1+\mathbf{XS})\right)$$
$$+ \int \bar{\Psi}^e e^{ie\chi}(\mathbf{X}+\mathbf{XSX})e^{-ie\chi}\Psi^e \tag{4.5}$$

where $\mathbf{X} \equiv e\partial\!\!\!/\chi$ and $\alpha_n \equiv 1+a_{\le n}$.

Note that (4.5) is similar to the identity (1.19) which was expressed in terms of another generating functional W. (4.5) is more complicated because of the inclusion of counterterms and because the spinor "sources" Ψ^e, $\bar{\Psi}^e$ are added to rather than multiplied against the spinor fields, as in the case of the sources η, $\bar{\eta}$ in the definition of W.

__Proof of Lemma 4.2:__ The change of variables $\Psi(x) = e^{ie\chi(x)}\Psi'(x)$, $\bar{\Psi} = e^{-ie\chi}\bar{\Psi}'$, $A = A'$ in the integrand of (4.3) gives

$$V_{e,n}(A^e, e^{ie\chi}\Psi^e, e^{-ie\chi}\bar{\Psi}^e) =$$
$$\log \int e^{(V+\delta V_{\le n})(\Phi'+\Phi^e) - a_{\le n}(\bar{\Psi}'+\bar{\Psi}^e)\mathbf{X}(\Psi'+\Psi^e)}\, e^{-\bar{\Psi}'\mathbf{X}\Psi'} dP(\Phi') + \text{const.} \tag{4.6}$$

Here we use an operator notation in lieu of integral signs.

We would like to make another change of variables of the form

$$\Psi' = \Psi'' + E\Psi^e,\ \bar{\Psi}' = \bar{\Psi}'' + \bar{\Psi}^e\bar{E},\ A' = A'' \tag{4.7}$$

chosen so that the $\bar{\Psi}'$, Ψ' terms in the exponent of the factor $e^{-\bar{\Psi}'\mathbf{X}\Psi'}dP(\Phi')$ can be rewritten

$$\bar{\Psi}'(\mathbf{X}+\mathbf{S}^{-1})\Psi' = \bar{\Psi}''\mathbf{S}^{-1}\Psi'' + (\bar{\Psi}'+\bar{\Psi}^e)\mathbf{X}(\Psi'+\Psi^e) - Q \tag{4.8}$$

where Q is some function of Ψ^e and $\bar{\Psi}^e$. For then the right side of (4.6) becomes

$$\log \int e^{(V+\delta V_{\leq n})(\Phi'+\Phi^e) - \alpha_n(\bar{\Psi}'+\bar{\Psi}^e)\mathbf{X}(\Psi'+\Psi^e)} dP(\Phi'') + Q(\Psi^e,\bar{\Psi}^e) + \text{const.} \tag{4.9}$$

which can be simply expressed in terms of $V_{e,n}$. To satisfy (4.8) we compute from (4.7) that

$$\Psi'(\mathbf{X}+\mathbf{S}^{-1})\Psi' - \bar{\Psi}''\mathbf{S}^{-1}\Psi'' - (\bar{\Psi}'+\bar{\Psi}^e)\mathbf{X}(\Psi'+\Psi^e)$$

$$= \bar{\Psi}^e(\bar{E}\mathbf{S}^{-1}-\mathbf{X})\Psi'' + \bar{\Psi}''(\mathbf{S}^{-1}E-\mathbf{X})\Psi^e + \bar{\Psi}^e(\bar{E}\mathbf{S}^{-1}E-\mathbf{X}-\bar{E}\mathbf{X}-\mathbf{X}E)\Psi^e. \tag{4.10}$$

If we choose $\bar{E} = \mathbf{XS}$ and $E = \mathbf{SX}$ then (4.10) $= -\bar{\Psi}^e(\mathbf{XSX}+\mathbf{X})\Psi^e$ so that

$$Q = \bar{\Psi}^e(\mathbf{XSX}+\mathbf{X})\Psi^e. \tag{4.11}$$

We return to equation (4.9), fill in Q, and make the substitution

$$(V+\delta V_{\leq n})(\Phi'+\Phi^e) - \alpha_n(\bar{\Psi}'+\bar{\Psi}^e)\mathbf{X}(\Psi'+\Psi^e)$$

$$= (V+\delta V_{\leq n})(A'+A^e+\alpha_{n-1}^{-1}\alpha_n\partial\chi, \Psi'+\Psi^e, \bar{\Psi}'+\bar{\Psi}^e) \tag{4.12}$$

in the exponent. With this we find

$$V_{e,n}(A^e, e^{ie\chi}\Psi^e, e^{-ie\chi}\bar{\Psi}^e) = V_{e,n}\big(A^e+\alpha_{n-1}^{-1}\alpha_n\partial\chi,\ (1+\mathbf{SX})\Psi^e,\ \bar{\Psi}^e(1+\mathbf{XS})\big) + \bar{\Psi}^e(\mathbf{XSX}+\mathbf{X})\Psi^e. \tag{4.13}$$

∎

According to (4.4) and (3.31) the counterterms of order e^{n+1} are defined in terms of the local part of $V_{0,n}(A^e,\psi_0^e,\bar{\psi}_0^e) \equiv V_{e,n}\Big|_{\bar{\psi}_i^e=\psi_i^e=0,\, i>0}$, that is the real field sector of $V_{e,n}$. Explicitly we have

$$LV_{0,n}(A,\psi,\bar{\psi}) = -e\int :\bar{\psi}A\!\!\!/\psi: + \{a_{n+1}\int :\bar{\psi}(-i\partial\!\!\!/)\psi: + b_{n+1}\int :\bar{\psi}\psi: + c_{n+1}\int :\bar{\psi}A\!\!\!/\psi:$$

$$+ d_{n+1}\int :F^2: + \lambda_{5,n+1}\int :(\partial\cdot A)^2: + \lambda_{6,n+1}\int :A^2: + \lambda_{7,n+1}\int :A^4:\} + \text{const.} + O(e^{n+2}). \tag{4.14}$$

A simpler Ward identity than (4.5) holds for the local part $LV_{0,n}$:

<u>Lemma 4.3:</u> Suppose $V+\delta V_{\leq n}$ has gauge invariant form. Then for any smooth polynomially bounded real function $\chi(x)$,

$$LV_{0,n}(A,\ \psi,\ \bar{\psi}) = (LV_{0,n})(A+\alpha_{n-1}^{-1}\alpha_n\partial\chi,\ e^{-ie\chi}\psi,\ e^{+ie\chi}\bar{\psi}) + e\int\bar{\psi}\not{\partial}\chi\psi + O(e^{n+2}). \tag{4.15}$$

To prove Lemma 4.3, we use a description of the localization operator in terms of the scaling properties of fields. Let the fields $\Phi \equiv (\Phi_1,\ldots,\Phi_N)$ have dimensions $(\delta_1,\ldots,\delta_N)$. A functional $W(\Phi)$ is said to have scaling dimension s if under the rescaling of the field

$$\Phi^\zeta(x)\equiv\left(\zeta^{\delta_1}\phi_1(\zeta x),\ldots,\zeta^{\delta_N}\phi_N(\zeta x)\right) \tag{4.16}$$

it satisfies

$$W(\Phi^\zeta) = \zeta^{-s}W(\Phi). \tag{4.17}$$

Formally, by a Taylor expansion of the fields, we express any effective potential $V(\Phi)$ as a sum

$$\sum_{s=-\infty}^{d} V^s(\Phi) \tag{4.18}$$

where each term V^s has scaling dimension s. We shall see that then the local part of V is given by the terms of non-negative dimension

$$LV(\Phi) \equiv \sum_{s\geq 0} V^s(\Phi)\ . \tag{4.19}$$

We have from (4.17) and (4.18)

$$\zeta^d V(\Phi^\zeta) = \sum_{s=-\infty}^{d} \zeta^{d-s}V^s(\Phi)\ .$$

Therefore,

$$\begin{aligned} V^s(\Phi) &= \frac{1}{(d-s)!}\,\partial_\zeta^{d-s}[\zeta^d V(\Phi^\zeta)]\big|_{\zeta=0}\ . \\ &\equiv T_s[V(\Phi^\zeta)]. \end{aligned} \tag{4.20}$$

While the sum (4.18) is formal, we may still make rigorous statements by taking (4.20) as the definition of V^s.

In particular, we may show that (4.19) agrees with the definition of L given in (2.48) by considering a typical monomial occurring in V,

$$W(\Phi) = \int w(x_1,\ldots,x_n):\Phi_{j_1}(x_1)\ldots\Phi_{j_n}(x_n):dx_1\ldots dx_n \,.$$

After a change of variables

$$x = \zeta x_1,\ z_2 = x_2-x_1,\ \ldots,\ z_n = x_n-x_1$$

we calculate using the translation invariance of w that

$$\zeta^d W(\Phi^\zeta) = \zeta^{\Sigma\delta_{j_i}} \int w(0,z_2,\ldots,z_n) \tag{4.21}$$

$$:\Phi_{j_1}(x)\Phi_{j_2}(x+\zeta z_2)\ldots\Phi_{j_n}(x+\zeta z_n):dx\, dz_2\ldots dz_n.$$

The right side of (4.21) is smooth in ζ for any fixed C_0^∞ fields Φ. Using (4.20) we find that the right side of (4.19) is given by

$$\int w(0,z_2,\ldots,z_n) \sum_{k=0}^{\delta} \frac{1}{k!}\,\delta_\zeta^k :\Phi_{j_1}(x)\Phi_{j_2}(x+\zeta z_2)\ldots\Phi_{j_n}(x+\zeta z_n):\Big|_{\zeta=0} dx\, dz_2\ldots dz_n \tag{4.22}$$

where $\delta = d-\Sigma\delta_{j_i}$. This agrees precisely with (2.48), and so we have established (4.19).

Some care should be used in applying the scaling dimension formulation (4.19)-(4.20) of the localization operator L. While it is true that $\zeta^d W(\zeta)$ is C^∞ in ζ (for Φ C_0^∞ and $w(x_1,\ldots,x_n)$ C^∞ and rapidly decaying except for translation invariance) one may not apply $T_s\cdot = \frac{1}{(d-s)!}\,\partial_\zeta^{d-s}\zeta^d\cdot\Big|_{\zeta=0}$ by evaluating the derivatives and then simply plugging $\zeta=0$ into the integrand. The resulting integral is $\int w(x_1,\ldots,x_n)dx_1\ldots dx_n$ which diverges by translation invariance. One should first make a change of variables like that used above (see (4.21)).

The advantage of (4.19) is that it allows an easy transfer of information from V to LV. We consider the effect of applying L to the Ward identity (4.5) restricted to $\psi_i=\bar\psi_i=0$, $i>0$. We specify the scaling behaviour of the right side of (4.5) by adopting, in addition to the rules

$$A^\zeta(x) \equiv \zeta A(\zeta x) \tag{4.22a}$$

$$\psi^\zeta(x) \equiv \zeta^{3/2}\psi(\zeta x),\ \bar\psi^\zeta(x) \equiv \zeta^{3/2}\bar\psi(\zeta x) \tag{4.22b}$$

the rule

$$\chi^\zeta(x) \equiv \chi(\zeta x). \tag{4.22c}$$

With this, $e^{-ie\chi}\psi$ scales like ψ and $\partial\chi$ scales like A. The spinor covariance, however, scales inhomogeneously. We calculate

$$\begin{aligned}\left[eS\partial\!\!\!/\chi e^{-ie\chi}\psi\right]^\zeta(x) &= e\int S(x,y)\ \partial\!\!\!/\chi^\zeta(y)e^{-ie\chi^\zeta(y)}\psi^\zeta(y)\ dy \\ &= e\zeta^{5/2}F(\zeta,\zeta x)\end{aligned} \tag{4.23a}$$

where

$$F(\zeta,x) = \int S(0,z)f(x+\zeta z)dz$$

and

$$f(x) = \partial\!\!\!/\chi(x)e^{-ie\chi(x)}\psi(x).$$

Similarly

$$[\bar\psi e^{ie\chi}e\partial\!\!\!/\chi S]^\zeta(x) = e\zeta^{5/2}G(\zeta,\zeta x) \tag{4.23b}$$

where

$$G(\zeta,x) = \int g(x+\zeta z)S(z,0)dz$$

and

$$g(x) = \bar\psi(x)e^{ie\chi(x)}\partial\!\!\!/\chi(x).$$

Now $F(\zeta,x)$ and $G(\zeta,x)$ are smooth near $\zeta=0$ for any x. The factor $e\zeta^{5/2}$ in (4.23) is to be compared with the factor $\zeta^{3/2}$ in (4.22b). It implies that if we take any monomial $W(\Phi^e)$ and substitute $e^{-ie\chi}\psi^e + SXe^{-ie\chi}\psi^e$ for Ψ^e (as happens in (4.5)) terms containing $SXe^{-ie\chi}\psi^e$ always have lower dimension than the corresponding terms containing $e^{-ie\chi}\psi^e$ and consequently are "less likely" to have a non-zero local part.

In general we have

<u>Lemma 4.4</u> If $\{Q_i^\zeta(x)\}$ are any C_0^∞ functions of the form

$$Q_i^\zeta(x) = \zeta^{\delta_i}Q_i(\zeta,\zeta x)$$

where δ_i is the dimension of Φ_i and $Q(\zeta,x)$ is smooth at $\zeta=0$ then.

$$T_s V(Q_i^\zeta) = T_s\left[(LV)(Q_i^\zeta)\right] .$$

Proof: Use k to denote d-s and suppress the subscript i. The left hand side is

$$T_s[V(Q^\zeta)] = \frac{1}{k!}\, \partial_\zeta^k [\tau^d V(\tau^\delta Q(\zeta,\tau x))|_{\tau=\zeta}]|_{\zeta=0}$$

$$= \sum_{\ell=0}^{k} \frac{1}{\ell!(k-\ell)!}\, \partial_\tau^\ell\, \partial_\zeta^{k-\ell}\, [\tau^d V(\tau^\delta Q(\zeta,\tau x))]|_{\zeta=\tau=0}.$$

Now

$$(LV)(\Phi) = \sum_{\ell=0}^{d} \frac{1}{\ell!}\, \partial_\sigma^\ell\, [\sigma^d V(\sigma^\delta \Phi(\sigma x))]|_{\sigma=0}.$$

Using this with $\Phi(x) = \zeta^\delta Q(\zeta,\zeta x)$ where ζ is fixed gives

$$T_s[LV(Q^\zeta)] = \frac{1}{k!}\, \partial_\zeta^k \{\zeta^d (LV)(\zeta^\delta Q(\zeta,\zeta x))\}|_{\zeta=0}$$

$$= \frac{1}{k!}\, \partial_\zeta^k \sum_{\ell=0}^{d} \{\frac{1}{\ell!}\, \partial_\sigma^\ell [(\zeta\sigma)^d V((\zeta\sigma)^\delta Q(\zeta,\sigma\zeta x))]\Big|_{\sigma=0}\}\Big|_{\zeta=0}$$

$$= \frac{1}{k!}\, \partial_\zeta^k \sum_{\ell=0}^{d} \{\frac{\zeta^\ell}{\ell!}\, \partial_\tau^\ell [\tau^d V(\tau^\delta Q(\zeta,\tau x))]\Big|_{\tau=0}\}\Big|_{\zeta=0}$$

where for fixed ζ, we have set $\tau = \zeta\sigma$. Now ℓ derivatives must hit ζ^ℓ, leaving $k-\ell$ derivatives. Hence

$$T_s[(LV)(Q^\zeta)] = \frac{1}{k!} \sum_{\ell=0}^{k} \frac{k!}{(k-\ell)!\,\ell!}\, \partial_\tau^\ell\, \partial_\zeta^{k-\ell}\, [\tau^d V(\tau^\delta Q(\zeta,\tau x))]\Big|_{\zeta=\tau=0}$$

$$= T_s[V(Q^\zeta)]. \qquad \blacksquare$$

Proof of Lemma 4.3: Substitute A^ζ, Ψ^ζ and χ^ζ into (4.5), set the fictitious fields to zero and apply T_s. By Lemma 4.4 we get

$$T_s V_{0,n}(A^\zeta,\psi^\zeta,\bar\psi^\zeta)$$

$$= T_s\{(LV_{0,n})(A^\zeta+\alpha_{n-1}^{-1}\alpha_n\partial(\chi^\zeta),\ e^{-ie\chi^\zeta}\psi^\zeta+e\zeta^{5/2}G(\zeta,\zeta x),\dots)\}$$

$$+ T_s\{\textstyle\int\bar\psi^\zeta e^{ie\chi^\zeta} X^\zeta e^{-ie\chi^\zeta}\psi^\zeta\} + T_s\{\textstyle\int\bar\psi^\zeta e^{ie\chi^\zeta} X^\zeta S X^\zeta e^{-ie\chi^\zeta}\psi^\zeta\}. \qquad (4.24)$$

Using formula (4.14) we see that the only terms of nonnegative dimension from $(LV_{0,n})(A^{\zeta}+\ldots,\ldots,\ldots)$ that contain $SX^{\zeta}e^{-ie\chi^{\zeta}}\psi^{\zeta}$ or $\bar{\psi}^{\zeta}e^{ie\chi^{\zeta}}X^{\zeta}S$ are

$$b_{n+1}\int\bar{\psi}^{\zeta}e^{ie\chi^{\zeta}}e\zeta^{5/2}F(\zeta,\zeta x)$$

and

$$b_{n+1}\int e\zeta^{5/2}G(\zeta,\zeta x)e^{ie\chi^{\zeta}}\psi^{\zeta} .$$

These are both $O(e^{n+2})$. Furthermore

$$T_s\{\int\bar{\psi}^{\zeta}e^{ie\chi^{\zeta}}X^{\zeta}e^{-ie\chi^{\zeta}}\psi^{\zeta}\} = \delta_{s,0}\int\bar{\psi}X\psi \tag{4.25}$$

and

$$T_s\{\int\bar{\psi}^{\zeta}e^{ie\chi^{\zeta}}X^{\zeta}SX^{\zeta}e^{-ie\chi^{\zeta}}\psi^{\zeta}\}$$

$$= T_s\{\zeta^5\int g(\zeta x)S(x-y)f(\zeta y)dxdy\}$$

$$= T_s\{\zeta\int g(y+\zeta z)S(z)f(y)dzdy\} = 0 \tag{4.26}$$

since $\int g(y+\zeta z)S(z)f(y)dzdy$ is regular at $\zeta=0$. So

$$T_sV_{0,n}(A^{\zeta},\psi^{\zeta},\bar{\psi}^{\zeta}) = T_s\{(LV_{0,n})\left(A^{\zeta}+\alpha_{n-1}^{-1}\alpha_n\partial(\chi^{\zeta}),\ e^{-ie\chi^{\zeta}}\psi^{\zeta},\ \bar{\psi}^{\zeta}e^{ie\chi^{\zeta}}\right)\}$$

$$+ \delta_{s,0}\int\bar{\psi}X\psi + O(e^{n+2})$$

and the lemma follows. ∎

<u>Proof of Theorem 4.1:</u> We write out the right side of (4.15) explicitly, using (4.14) and the fact that $\alpha_{n-1}^{-1}\alpha_n = 1 + a_n + O(e^{n+1})$:

$$(-e+c_{n+1})\int:\bar{\psi}[\not{A}+(1+a_n)\not{\partial}\chi]\psi: + a_{n+1}\int:\bar{\psi}(-i\not{\partial})\psi:$$

$$+ b_{n+1}\int:\bar{\psi}\psi: + d_{n+1}\int:F^2: + \lambda_{5,n+1}\int:[\partial\cdot(A+\partial\chi)]^2:$$

$$+ \lambda_{6,n+1}\int:(A+\partial\chi)^2: + \lambda_{7,n+1}\int:(A+\partial\chi)^4:$$

$$+ e\int:\bar{\psi}\not{\partial}\chi\psi: + \text{const.} + O(e^{n+2}). \tag{4.27}$$

This is χ-independent from which it follows that

$$\lambda_{5,n+1} = \lambda_{6,n+1} = \lambda_{7,n+1} = 0 \qquad (4.28)$$

and

$$0 = (-e+c_{n+1})(1+a_n) + e + O(e^{n+2})$$

$$= c_{n+1} - ea_n \ . \qquad (4.29)$$

This verifies both conditions (4.2a) and (4.2b) for δV_{n+1}, under the hypothesis that $\delta V_{\leq n}$ has gauge invariant form, when I=0.

We shall outline the extension of this argument to the case I<0 in §7. ∎

§5. The Limits $\Lambda \to \infty$ and $U \to \infty$

We consider the renormalized tree expansion for loop regularized QED in the presence of the photon cutoffs U and I = 0. We shall show that for the effective potential evaluated with external fictitious fields set equal to zero, the limit $\Lambda\to\infty$ exists. In fact, any graph with no (external) fictitious field (ff) legs but with one or more (internal) ff loops actually vanishes in this limit. All dependence on fictitious fields and the problem of wrong counterterms drop out, after the fictitious fields have performed their crucial role of providing finite gauge invariant counterterms. When $\Lambda\to\infty$, the resulting effective potential $V^{I,U}$ is fully renormalized according to the rules of §2, and Theorem 2.5 applies to prove convergence of the $U\to\infty$ limit. The main result of this section is:

__Theorem 5.1.__ Let G^Λ be a graph contributing to $V_r^{I,U,\Lambda}$. If G^Λ has at least one loop of ff lines (possibly hidden inside counterterms), but no ff legs, then

$$\lim_{\Lambda\to\infty} \| G^\Lambda \| = 0. \tag{5.1}$$

Since a graph without ff loops or ff legs is completely independent of Λ, we have:

__Corollary 5.2.__ For $-1 \leq r \leq U$

$$V_r^{I,U}(A,\psi_0,\bar{\psi}_0) \equiv \lim_{\Lambda\to\infty} V_r^{I,U,\Lambda}\Big|_{\psi_i=\bar{\psi}_i=0,\ i>0} \tag{5.2}$$

exists in perturbation theory.

The effective potentials $V_r^{I,U}$ are given by a renormalized tree expansion which contains no residue of the fictitious fields and consists exactly of the contributions from graphs built of real field (rf) lines (i.e. electron and photon lines), renormalized according to the rules of §2. The electron propagator has no cutoff, while the photon propagator still has the cutoffs U and I = 0. But now, since $V_r^{I,U}$ is fully UV renormalized, Theorem 2.5 applies and gives bounds on $V_r^{I,U}$ which are uniform in U. In this way we immediately obtain:

<u>Theorem 5.3.</u> QED_4 is UV-renormalizable using gauge invariant counterterms. That is, for $-1 \le r < \infty$,

$$V_r^I \equiv \lim_{U\to\infty} V_r^{I,U} \tag{5.3}$$

exists in perturbation theory. The contribution to V_r^I from a graph G satisfies the bounds of Theorem 2.5.

Photon lines are harmless in the presence of the cutoff U. They obey the bounds

$$|\partial_x^n D^{(h)}(x,y)| < c\, M^{(2+|n|)U} e^{-M^h|x-y|}. \tag{5.4}$$

As a result, ff graphs (i.e., graphs with external ff legs) that would normally require renormalization in fact do not, and we may choose "wrong" counterterms for these graphs (as we were obliged to do in order to maintain loop regularization in §3). Consider for example the graph

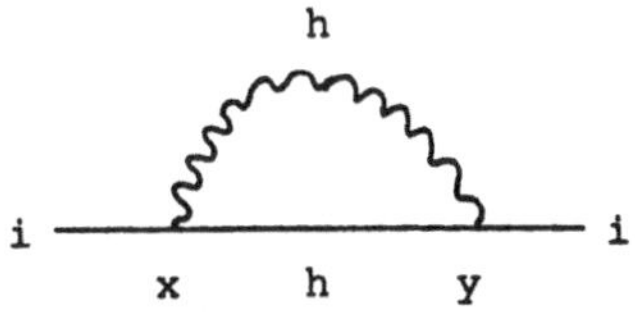

(5.5)

whose kernel is $G^{(h)}(x,y) = D^{(h)}(x,y)\ \gamma^\mu S_i^{(h)}(x,y)\ \gamma^\mu$ (μ summed over). By (5.4) and (1.15a),

$$\sum_{h=0}^{\infty} \int |G^{(h)}(x,y)|dy \le c_1 \sum_{h=0}^{\infty} M^{2U+3h-4h} \le c_2 M^{2U} \tag{5.6}$$

so that (5.5) does not require renormalization so long as $U < \infty$. Moreover, (5.6) $\to 0$ when $\Lambda \to \infty$. To see this, we bound

$$S_i^{(h)}(x,y) = (2\pi)^{-2}(i\partial\!\!\!/_x + M_i) \int_{M^{-2h}}^{M^{-2h+2}} e^{-\alpha M_i^2 - (x-y)^2/4\alpha}\ \alpha^{-2} d\alpha$$

as in (1.15a) except that instead of estimating the factor $e^{-\alpha M_i^2}$ by 1 we estimate it by

$$c_3 \alpha^{-a/2} M_i^{-a}$$

for $a \geq 0$. In this way we obtain

$$|\partial_x^n S_i^{(h)}(x,y)| \leq c_4 \Lambda^{-a} M^{(3+a+|n|)h} e^{-M^h|x-y|} \tag{5.7}$$

for any $a \geq 0$. For $0 < a < 1$ (5.6) is replaced by

$$\sum_{h=0}^{\infty} \int |G^{(h)}(x,y)|dy \leq c_5 \Lambda^{-a} M^{2U} \sum_{h=0}^{\infty} M^{(a-1)h} \leq c_6(U)\Lambda^{-a} \to 0$$

as $\Lambda \to \infty$. It is the burden of this section to show that these properties of example (5.5)(namely that it vanishes when $\Lambda \to \infty$ even without renormalization) are shared by all graphs with ff dependence.

It is convenient to reorganize the loop regularized version of the renormalized tree expansion. While the counterterms as defined in terms of the operator L^Λ (see equation (3.31)) are needed to preserve loop regularization, the corresponding renormalization operator R is no help at all for ff graphs. A general tree can be written as a sum

by attaching labels r and f to the lowest fork to distinguish the contributions from rf graphs (i.e. graphs without external ff legs) and ff graphs. We redefine R and C applied to an f-fork: R is now taken to have no effect

, (5.8)

while C provides all the ff counterterms we need

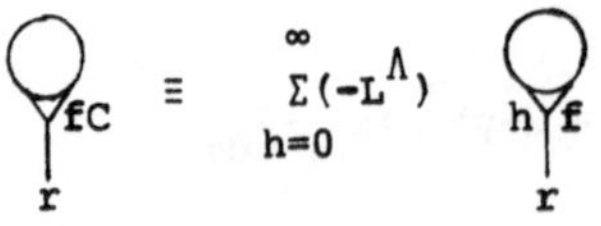

. (5.9)

R and C act on r-forks in the usual way. In our modified tree expansion for $V_r^{I,U,\Lambda}$ the sum over R and C labellings is replaced by a sum over fork labellings with labels taken from the set {fR, fC, rR, rC}. To see that this is legitimate, we note that

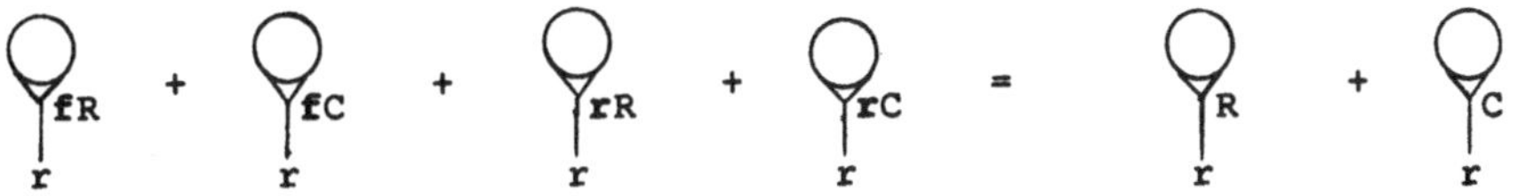

$$\bigcirc_{\substack{fR \\ r}} + \bigcirc_{\substack{fC \\ r}} + \bigcirc_{\substack{rR \\ r}} + \bigcirc_{\substack{rC \\ r}} = \bigcirc_{\substack{R \\ r}} + \bigcirc_{\substack{C \\ r}} \quad . \tag{5.10}$$

We denote by $\mathfrak{J}_{fR}(\tau)$ the set of all forks of τ with label fR, and similarly

for other labellings. In particular,

$$\mathfrak{J}_R = \mathfrak{J}_{fR} \cup \mathfrak{J}_{rR} \ , \quad \mathfrak{J}_C = \mathfrak{J}_{fC} \cup \mathfrak{J}_{rC} \quad .$$

Proof of Theorem 5.1. We first introduce some additional bookkeeping notation for

graphs G:

$$p(G) \equiv \sum_{\ell \in \mathcal{L}_p(G)} d_\ell \tag{5.11}$$

where $\mathcal{L}_p(G) \equiv \{\text{photon lines of } G\}$ and

$$\phi(G) \equiv \text{number of ff loops of G.} \tag{5.12}$$

For labelled subgraphs G_f

$$P_f \equiv p(G_f/\{\text{C-subgraphs of } G_f\}) \tag{5.13a}$$

$$p_f \equiv p(g_f) \tag{5.13b}$$

$$\Phi_f \equiv \phi(G_f/\{\text{fR and fC subgraphs of } G_f\}) \tag{5.14a}$$

$$\phi_f \equiv \phi(g_f) \tag{5.14b}$$

$\left(\text{recall } g_f \equiv G_f/\{G_{f'} : \pi(f') = f\}\right)$. Note the formulas

$$P_f = p_f + \sum_{\substack{f' \in \mathfrak{J}_{rR} \cup \mathfrak{J}_{fR} \\ \pi(f')=f}} P_{f'} \ , \tag{5.15}$$

$$\Phi_f = \phi_f + \sum_{\substack{f' \in \mathfrak{J}_{rR} \cup \mathfrak{J}_{rC} \\ \pi(f')=f}} \Phi_{f'} \quad . \tag{5.16}$$

We base our proof on the following modification of the bound (2.86)

$$\|G\| \leq c \sum_{\vec{h} \in \tilde{\mathcal{H}}(\tau,\vec{\rho})} \prod_{f \in \mathfrak{J}} \left\lceil M^{\delta_f(h_f - h_{\pi(f)})} \, M^{Up_f} \, M^{-P_f h_f} \right\rceil \prod_{\substack{f \in \mathfrak{J}_{fC} \\ \delta_f \geq 0}} \Lambda^{\delta_f} \Lambda^{-m_f} \tag{5.17}$$

where $c = c(G)$ will denote various constants independent of $\vec{h}$ and Λ

and $m_f \equiv$ number of ff mass counterterms on external lines of G_f

but not in any C-subgraph.

To justify (5.17), note first that since photon lines are here bounded by (5.4) we have improvement factors $M^{Up_f} M^{-p_f h_f}$ at each fork. The factors $\Lambda^{\delta_f} \Lambda^{-m_f}$ at each $f \in \mathfrak{F}_{fC}$ arise from the definition of L^Λ. Noting that $\Lambda < M_i(\Lambda) < 2\Lambda$ for $i > 0$, we can see from (3.31) that a factor Λ^{δ_f} occurs when g_f is a ff mass counterterm, while a factor Λ^{-m_f} always arises from the cancellation of Λ factors (from ff mass counterterms in G_f) resulting from the replacement operation:

ff external line ⟶ rf external line.

In view of (5.8) it is clear that the bound (5.17) cannot include a renormalization improvement at forks $f \in \mathfrak{F}_{fR}$. In particular, the rule for choosing $D(G_f)$ and N_f adopted in §2 must be modified here by requiring that $n_f = 0$ for $f \in \mathfrak{F}_{fR}$. Thus, while we still have $\delta_f < 0$ for $f \in \mathfrak{F}_{fR}$ and $\delta_f \geq 0$ for $f \in \mathfrak{F}_C$, δ_f has no definite sign for $f \in \mathfrak{F}_{fR}$. We also remark that (5.8b) requires that the range of scales $\tilde{\mathcal{H}}(\tau,\vec{\rho})$ in (5.17) differs from $\mathcal{H}(\tau,\vec{\rho})$ given by (2.50), in that at an fC-fork, h_f ranges over 0, 1, ... instead of 0, 1, ..., $h_{\pi(f)}$.

For any ff line of scale h, we may introduce a factor

$$\Lambda^{-a} M^{ah} \tag{5.18}$$

for any value $a \geq 0$, by use of the estimate (5.7). Very roughly, our plan is to introduce a suitably balanced collection of factors (5.18) which will overcome the positive powers of Λ accrued through the L^Λ operation, but which will not upset the convergence of the sums over scales.

We will use the inductive method of §2 [see (2.102) - (2.105)] to bound the contribution to $\|G\|$ coming from a labelled or unlabelled subgraph G_f. Here we define the contribution at an unlabelled fork f to be [c.f.(2.103)]

$$U_f(h_f) = \sum_{\substack{\vec{h} \in \mathcal{H}_f \\ h_f \text{ fixed}}} \prod_{f'>f} \left[M^{\delta_{f'}(h_{f'}-h_{\pi(f')})} M^{-p_{f'} h_{f'}} \right] \prod_{\substack{f'>f \\ f' \in \mathfrak{F}_{fC} \\ \delta_{f'} \geq 0}} \Lambda^{\delta_{f'}-m_{f'}} \tag{5.19}$$

The contribution $\bar{U}_f(k)$ at a labelled fork f with $h_{\pi(f)} = k$ is given by

$$\bar{U}_f(k) = \begin{cases} \sum_{h_f=0}^{\infty} \Lambda^{\delta_f - m_f} M^{\delta_f(h_f-k)} U_f(h_f) & \text{if } f \in \mathcal{J}_{fC},\ \delta_f \geq 0 \\ 0 & \text{if } f \in \mathcal{J}_{fC},\ \delta_f < 0 \end{cases} \tag{5.20a}$$

$$\bar{U}_f(k) = \begin{cases} \sum_{h_f \leq k} M^{\delta_f(h_f-k)} U_f(h_f) & \text{if } f \in \mathcal{J}_{rC},\ \delta_f \geq 0 \\ 0 & \text{if } f \in \mathcal{J}_{rC},\ \delta_f < 0 \end{cases} \tag{5.20b}$$

$$\bar{U}_f(k) = \sum_{h_f > k} M^{\delta_f(h_f-k)} U_f(h_f) \qquad \text{if } f \in \mathcal{J}_R\,. \tag{5.20c}$$

By making an inductive hypothesis on the contributions U_f we can calculate bounds on the $\bar{U}_f$'s, then use the formula

$$U_f(h_f) = M^{-p_f h_f} \prod_{f':\pi(f')=f} \bar{U}_{f'}(h_f)$$

to verify the inductive hypothesis.

The inductive hypothesis for the contribution U_f from an unlabelled fork f is

$$U_f(h_f) < c(h_f+1)^{\kappa_f} \Lambda^{-b\Phi_f + m_f} M^{(-p_f - m_f + b\Phi_f)h_f}, \tag{5.21}$$

for all b satisfying $0 \leq b(\max_f \Phi_f) < 1$ where κ_f is the number of C-forks f'>f. The number c is independent of h_f and Λ.

Note that when G_f has no ff dependence, the bound reduces to

$$c\ (h_f+1)^{\kappa_f} M^{-p_f h_f}$$

which is just (2.107a) modified by the removal of factors from photon lines.

<u>Lemma 5.4.</u> Suppose the inductive hypothesis (5.21) holds for the contribution from a fork f. Then $\bar{U}_f(k)$ is bounded by

$$c\Lambda^{\delta_f} M^{-\delta_f k} \qquad \text{if } f \in \mathcal{J}_{fC}\,; \tag{5.22a}$$

$$c(k+1)^{\kappa_f} \Lambda^{m_f} M^{(-p_f - m_f)k} \qquad \text{if } f \in \mathcal{J}_{fR}\,; \tag{5.22b}$$

$$c(k+1)^{\bar{\kappa}_f}\,\Lambda^{-b\Phi_f}\,M^{b\Phi_f k} \qquad \text{if } f\in\mathcal{F}_{rC}\,; \tag{5.22c}$$

$$c(k+1)^{\kappa_f}\,\Lambda^{-b\Phi_f}\,M^{(-P_f+b\Phi_f)k} \qquad \text{if } f\in\mathcal{F}_{rR}\,. \tag{5.22d}$$

<u>Proof</u>. If $f\in\mathcal{F}_{fC}$, then by (5.20a) and (5.21) with $b = 0$ we have

$$\bar{U}_f(k) \le c\,\Lambda^{\delta_f} M^{-\delta_f k} \sum_{h_f=0}^{\infty} (h_f+1)^{\kappa_f} M^{a_f h_f}$$

where $a_f = \delta_f - P_f - m_f$. Obviously (5.22a) holds when $a_f < 0$. The exceptional cases, with $a_f = 0$ or 1 are:

$$\delta_f = 1,\; P_f = 0,\; m_f = 0 \tag{5.23a}$$

$$\delta_f = 1,\; P_f = 0,\; m_f = 1 \tag{5.23b}$$

$$\delta_f = 0,\; P_f = 0,\; m_f = 0\;. \tag{5.23c}$$

In these three cases, we claim that the counterterms of degree δ_f (and $\delta_f - a_f$) vanish identically and so (5.22a) holds trivially. For $P_f = 0$ implies that the graph G_f consists of a line of $\bar{\psi}\psi$, $\bar{\psi}(-i\not{\partial})\psi$ or $\bar{\psi}\not{A}\psi$ vertices. For example, a case b) graph has wave function vertices and a single mass vertex (the slash denotes differentiation):

wave function vertices mass vertex

L_1 applied to such graphs gives 0.

Suppose $f\in\mathcal{F}_{fR}$. Then from (5.20c) and (5.21) with $b = 0$ we have

$$\bar{U}_f(k) \le c\Lambda^{m_f} M^{-\delta_f k} \sum_{h_f > k} (h_f+1)^{\kappa_f} M^{a_f h_f}.$$

Except for the three classes of graphs (5.23) the exponent $a_f \le -1$ and we have the desired bound. But we have just seen that L_{δ_f} and $L_{\delta_f - a_f}$ applied to these exceptional graphs give zero. This means we can renormalize these graphs for free with these vanishing counterterms, gaining the improvement factor

$$M^{-(a_f+1)(h_f-k)}\;.$$

Then the sums over scales converge and yield in each case the bound (5.20b).

For $f \in \mathcal{J}_{rC}$ with $\delta_f \geq 0$ we have

$$\bar{U}_f(k) \leq c\Lambda^{-b\Phi_f} \sum_{h_f \leq k} (h_f+1)^{\kappa_f} M^{\delta_f(h_f-k)} M^{(-P_f+b\Phi_f)h_f}$$

$$\leq c\Lambda^{-b\Phi_f}(k+1)^{\bar{\kappa}_f} M^{b\Phi_f k} .$$

Finally, for $f \in \mathcal{J}_{rR}$

$$\bar{U}_f(k) \leq c\Lambda^{-b\Phi_f} \sum_{h_f > k} (h_f+1)^{\kappa_f} M^{\delta_f(h_f-k)} M^{(-P_f+b\Phi_f)h_f}$$

The sum over h_f always converges if $b\Phi_f < 1$, and gives the bound (5.22d).

∎

Proof of Theorem 5.1 (continued). We first verify the inductive hypothesis. If $f_1,\ldots,f_q$ are the labelled forks immediately above the fork f (there are also s-q endpoints feeding into f), then (see (2.104))

$$U_f(h_f) = M^{-p_f h_f} \prod_{i=1}^{q} \bar{U}_{f_i}(h_f) . \tag{5.24}$$

By the bounds of Lemma 5.4 we have

$$U_f(h_f) \leq c(h_f+1)^{\kappa_f} \Lambda^{\tilde{\Delta}(G_f)} M^{\tilde{\Gamma}(G_f)h_f} \tag{5.25}$$

where

$$\tilde{\Delta}(G_f) \equiv -b \sum_{f_i \in \mathcal{J}_r} \Phi_{f_i} + \sum_{f_i \in \mathcal{J}_{fR}} m_{f_i} + \sum_{f_i \in \mathcal{J}_{fC}} \delta_{f_i}$$

$$\tilde{\Gamma}(G_f) = -p_f - \sum_{f_i \in \mathcal{J}_{fC}} \delta_{f_i} - \sum_{f_i \in \mathcal{J}_R} P_{f_i} - \sum_{f_i \in \mathcal{J}_{fR}} m_{f_i} + b \sum_{f_i \in \mathcal{J}_r} \Phi_{f_i}$$

$$\kappa_f = \sum_{f_i} \bar{\kappa}_{f_i} .$$

Using formulas (5.15) and (5.16) we obtain

$$\tilde{\Delta}(G_f) = -b\Phi_f + m_f + E_f \tag{5.26}$$

$$\tilde{\Gamma}(G_f) = -P_f - m_f + b\Phi_f - E_f \tag{5.27}$$

where

$$E_f \equiv b\phi_f + \sum_{f_i \in \mathfrak{I}_{fC}} \delta_{f_i} - m_f + \sum_{f_i \in \mathfrak{I}_{fR}} m_{f_i} . \qquad (5.28)$$

Now

$$m_f \leq \sum_{f_i \in \mathfrak{I}_{fR}} m_{f_i} + \sum_{f_i \in \mathfrak{I}_{fC}} \delta_{f_i}$$

with equality if the reduced graph g_f has no ff loops. It follows that $E_f=0$ when $\phi_f=0$, and in this case the bound (5.21) is verified.

If $\phi_f > 0$ then $E_f > 0$. In this case, g_f has at least one ff line, which may be hard or soft. For this particular line, we exercise our option to introduce a factor (5.18) into the main bound, choosing the value $a = E_f$. With this choice, we again obtain the bound (5.21).

The value of G^Λ has the bound (k is the root scale)

$$\|G^\Lambda\| \leq c\ M^{\delta_F k}\ \bar{U}_F(k)$$

to which we apply (5.22c) or (5.22d). G^Λ has ff loops and no ff legs so $\phi_f \geq 1$ and $f \in \mathfrak{I}_{rR} \cup \mathfrak{I}_{rC}$. By choosing $b>0$ we can conclude that $\|G^\Lambda\| \to 0$ as $\Lambda \to \infty$. ∎

§6. The Tree Expansion in the Infrared Regime

In this section we develop a tree expansion for a general field theory involving massless fields. Since not much renormalization is required in the infrared regime, there is considerable freedom in how we define the R- and C-operations and in how we organize the "spring" factors in the bounds. We shall exploit this freedom to produce a spring, $M^{b_f\left(h_f-h_{\pi(f)}\right)}$, at each R-fork f whose "strength" b_f is not only negative but grows linearly in $\delta(G_f)$.

The decomposition of a massless field involves negative as well as positive scales: the covariance with UV cutoff $U \geq 0$ and IR cutoff $I \leq 0$ is $C^{[I,U]} = \sum_{h=I}^{U} C^{(h)}$. We assume that all nonzero masses are at least 1 (scale if necessary) so that $C_{ij}^{(h)} \equiv 0$ for $h < 0$ if the fields Φ_i and Φ_j are massive. We still have the basic bound

$$|\partial_x^n C_{ij}^{(h)}(x,y)| \leq c\, M^{(2\delta_i+|n|)h}\, e^{-M^h|x-y|} \tag{6.1}$$

but the power counting roles of $M^{(2\delta_i+|n|)h}$ and $e^{-M^h|x-y|}$ are reversed when h changes sign. When $h < 0$ each line $C^{(h)}$ produces a good factor $M^{2\delta_i h}$ and each derivative ∂_x acting on it produces an additional good factor M^h. On the other hand each vertex integrated over $\mathbb{R}^d$ at scale h produces a bad factor M^{-dh}. Consequently interaction vertices v with degree $\delta(v) < 0$ $\big($see (2.84)$\big)$ cause problems in the UV regime ("nonrenormalizability") but reduce the need for renormalization in the IR regime ("superrenormalizability"). Conversely when $\delta(v) > 0$, v is UV superrenormalizable but IR nonrenormalizable (if all fields at the vertex are massless). Later in this section we shall assume that the interaction V is dimensionless i.e. every vertex v satisfies $\delta(v) = 0$.

A second consequence of the above role reversal of the factors M^{-dh} is that it becomes important at negative scales to distinguish between internal and external vertices of a graph G. The former involve integrations of propagators over $\mathbb{R}^d$ and lead to bad factors M^{-dh}. The latter involve integrations against test functions Φ^e. If the test functions are chosen to be supported in unit balls (or at least to be integrable) the corresponding M^{-dh}'s disappear. In momentum space language, we avoid evaluating G at exceptional momenta by averaging. This is, of course, one reason for selecting the norm (2.21). For earlier work on IR power counting, see References 26 and 27 and references cited therein.

Recall that the reduced graph $g_f = G_f/\{G_{f'}: \pi(f') = f\}$ has ordinary vertices and generalized vertices $G_{f'}$ (with $\pi(f') = f$). Here too an ordinary vertex is external if it has an external field Φ^e attached to it and a generalized vertex $G_{f'}$ is external if $G_{f'}$ itself contains at least one ordinary external vertex. For a subgraph H of G, we write

$$\mathcal{V}^e(H) = \text{set of external vertices of } H$$

and

$$v^e(H) = |\mathcal{V}^e(H)| \ .$$

If $\lambda \in \Lambda(H)$ is a leg (= half-line or field) of H we say it is external if it corresponds to an external field Φ^e and internal if it corresponds to a field $\Phi^{(\leq k)}$, and we write

$$\Lambda(H) = \Lambda^e(H) \cup \Lambda^i(H) \equiv [\Lambda(H) \cap \Lambda(G)] \cup [\Lambda(H) \setminus \Lambda(G)] \ . \tag{6.2a}$$

If $\Delta(H) = \sum_{\lambda \in \Lambda(H)} \delta_\lambda$ is the total dimension of the legs of H (counting derivatives) we then have

$$\Delta(H) = \Delta^e(H) + \Delta^i(H) \equiv [\sum_{\lambda \in \Lambda^e(H)} \delta_\lambda] + [\sum_{\lambda \in \Lambda^i(H)} \delta_\lambda] \ . \tag{6.2b}$$

We similarly decompose $\Lambda(v)$ and $\Delta(v)$ for an interaction vertex v.

In our tree expansion we make the corresponding distinction by writing

$$V(\Phi + \Phi^e) = [V(\Phi + \Phi^e) - V(\Phi)] + V(\Phi)$$

at each endpoint of a tree. An endpoint is external if the first term is chosen and internal if the second is chosen; our tree expansions will include a sum over external/internal labels on the endpoints. We call a fork external (internal) if it has at least one (no) external endpoint above it, and we write $\mathcal{F}_e(\mathcal{F}_i)$ for the set of external (internal) forks.

The unrenormalized tree expansion is then unchanged from the UV case (see

(2.18a)) except that the scales may now take negative values if $r < -1$ and there are e/i labels on the endpoints. To generalize the bounds of §2 we shall need some new degrees of divergence. As always we have

$$D(H) = \sum_{\ell \in \mathcal{L}(H)} d_\ell - d\big(v(H) - 1\big) \tag{6.3a}$$

and

$$\delta(H) = d - \sum_{\lambda \in \Lambda(H)} \delta_\lambda \tag{6.3b}$$

with the former defined in terms of the internal lines of H and the latter defined in terms of the external legs of H. In the infrared regime we must also take into account the external vertices of H:

$$D(H,h) = D(H) + \chi(h<0)\, E(H) \tag{6.4a}$$

where

$$E(H) = d\big(v^e(H) - 1\big)_+ \tag{6.4b}$$

and $(x)_+ = \max(x,0)$ and $(x)_- = \min(x,0)$;

$$\delta(H,h) = \begin{cases} \delta(H) & \text{if } h \geq 0 \text{ or } v^e(H) = 0 \\ -\Delta^i(H) & \text{if } h < 0 \text{ and } v^e(H) > 0. \end{cases} \tag{6.4c}$$

We collect the relationships among all these degrees in:

<u>Lemma 6.1.</u> a) $D(G,h) = D(G)$ and $\delta(G,h) = \delta(G)$ if $h \geq 0$ or $v^e(G) = 0$.

b) $$D(G) + \sum_{v \in \mathcal{V}(G)} \delta(v) = \delta(G) \,. \tag{6.5a}$$

If V is dimensionless, $D(G) = \delta(G)$.

c) $$D(G_f,h) + \sum_{v \in \mathcal{V}(G_f)} \delta(v) = \delta(G_f,h) + \chi(h<0)\, B_f \tag{6.5b}$$

where

$$B_f = \sum_{v \in \mathcal{V}^e(G_f)} \big(d - \Delta^e(v)\big) = \sum_{f' \geq f} b_{f'}$$

and

$$b_f = \sum_{\substack{v \in \mathcal{V}^e(G) \\ \pi(v) = f}} \big(d - \Delta^e(v)\big) \,. \tag{6.5c}$$

If V is dimensionless,

$$D(G_f,h) = \delta(G_f,h) + \chi(h<0)\; B_f \tag{6.5d}$$

where

$$B_f = -\sum_{v\in\mathcal{V}^e(G_f)} \Delta^i(v) \; .$$

d) $$D(g_f) + \sum_{\pi(f_j)=f} \delta(G_{f_j}) + \sum_{v\in\mathcal{V}(g_f)} \delta(v) = \delta(G_f) \tag{6.6a}$$

where the first sum runs over the generalized (not ordinary) vertices of g_f and the second sum over the ordinary vertices of g_f.

e) $$D(g_f,h) + \sum_{\pi(f_j)=f} \delta(G_{f_j}) + \sum_{v\in\mathcal{V}(g_f)} \delta(v) = \delta(G_f,h) + \chi(h<0)\; B_f \; . \tag{6.6b}$$

Proof. a) Immediate from the definitions.

b) If G has n vertices

$$\begin{aligned} D(G) &= \sum_{\ell\in\mathcal{L}(G)} d_\ell - d(n-1) \\ &= \sum_v \sum_{\lambda\in\Lambda(v)} \delta_\lambda - \sum_{\lambda\in\Lambda(G)} \delta_\lambda - dn + d \\ &= \sum_v \Big(\sum_{\lambda\in\Lambda(v)} \delta_\lambda - d\Big) + \delta(G) \\ &= \delta(G) - \sum_v \delta(v) \; . \end{aligned}$$

c) (6.5b) reduces to (6.5a) except when $h < 0$ and $v^e(G_f) > 0$. In this case

$$\begin{aligned} D(G_f,h) + \sum_{v\in\mathcal{V}(G)} \delta(v) - \delta(G_f,h) &= \delta(G_f) + d\big(v^e(G_f)-1\big) + \Delta^i(G_f) \\ &= \delta(G_f) + dv^e(G_f) - d + \sum_{\lambda\in\Lambda(G_f)} \delta_\lambda - \sum_{v\in\mathcal{V}^e(G_f)} \sum_{\lambda\in\Lambda^e(v)} \delta_\lambda \\ &= \sum_{v\in\mathcal{V}^e(G_f)} \big(d-\Delta^e(v)\big) = B_f \; . \end{aligned}$$

d) The proof is the same as that of (6.5a).

e) The proof is the same as that of (6.5b).

∎

We consider now an unrenormalized graph G associated with a tree τ for which some of the vertices of G may be dimensionful and some of the scales $\vec{h}$ of τ may be negative. We generalize the bound of Theorem 2.2 as follows:

Theorem 6.2. For a graph G as described above, we have the following bounds on the norm (2.21) of its kernel:

a)
$$\|G\| \le K^{\ell(G)} \prod_{f\in\mathfrak{F}} M^{h_f D(g_f,h_f)} \tag{6.7a}$$

$$\le K^{\ell(G)} \prod_{f\in\mathfrak{F}} M^{h_f D(g_f)} . \tag{6.7b}$$

b)
$$\|G\| \le K^{\ell(G)} \prod_{f\in\mathfrak{F}} M^{h_f D(G_f,h_f) - h_{\pi(f)} D(G_f,h_{\pi(f)})} \tag{6.7c}$$

$$\le K^{\ell(G)} \prod_{f\in\mathfrak{F}} M^{(h_f - h_{\pi(f)}) D(G_f)} \tag{6.7d}$$

where $h_{\pi(F)} = 0$ for the lowest fork F of τ.

c) If V is dimensionless then:

$$\|G\| \le K^{\ell(G)} \prod_{f\in\mathfrak{F}} M^{(h_f - h_{\pi(f)}) \delta(G_f)} \tag{6.7e}$$

d) Let $\mathcal{V}^e_0(G) = \{v \in \mathcal{V}^e(G) \,|\, h_{\pi(v)} < 0\}$.

$$\|G\| \le K^{\ell(G)} \prod_{f\in\mathfrak{F}} M^{h_f \delta(G_f,h_f) - h_{\pi(f)} \delta(G_f,h_{\pi(f)})} \prod_{v\in\mathcal{V}(G)} M^{-h_{\pi(v)} \delta(v)}$$

$$\cdot \prod_{v\in\mathcal{V}^e_0(G)} M^{h_{\pi(v)} (d-\Delta^e(v))} . \tag{6.8}$$

<u>Proof</u> a) The proof is very similar to (2.25) - (2.28) except that we drop a certain number of lines from the integration tree $\mathcal{M}$. The vertices of g_f are connected by the $v(g_f) - 1$ hard lines of $\mathcal{M}_f = \mathcal{M} \cap g_f$ (see (2.24)) and these lines supply the factor (see (2.25))

$$\prod_{\ell\in\mathcal{M}_f} e^{-M^{h_f}|\ell|} \tag{6.9}$$

Suppose that f is a fork of τ with $h_f < 0$. If g_f has $v^e(g_f) > 1$ external vertices, we remove $v^e(g_f)-1$ lines from $\mathcal{M}_f$ to give a set of $v^i(g_f)$ lines $\tilde{\mathcal{M}}_f$ which connect the vertices of g_f into $v^e(g_f)$ separate connected pieces, each having exactly one external vertex. For example:

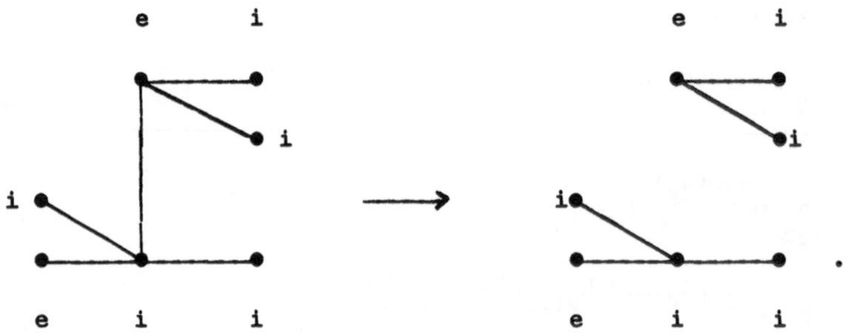

$\mathcal{M}_f$ $\qquad\qquad$ $\tilde{\mathcal{M}}_f$

If $v^e(g_f) \leq 1$, we remove no lines. Thus the number of lines removed is $(v^e(g_f)-1)_+$. Dropping the associated decay factors of these lines from (6.9), we bound (6.9) by $\prod_{\ell \in \tilde{\mathcal{M}}_f} e^{-M^{h_f}|\ell|}$ and then integrate out vertices in each piece just as we did for (2.25). This procedure gives an improvement of $M^{dh_f(v^e(g_f)-1)_+}$ over the estimate $M^{h_f D(g_f)}$ obtained previously. So (6.7a) follows. (6.7b) follows immediately from the fact that $D(g_f,h_f) \geq D(g_f)$, the inequality being strict only when $h_f < 0$ (and $v^e(g_f) > 1$).

b) Let $e_f = E(g_f)$ and $E_f = E(G_f)$. Then $E_f = \sum_{f' \geq f} e_{f'}$. By (6.4a)

$$\sum_{f \in \mathfrak{F}} [h_f D(G_f,h_f) - h_{\pi(f)} D(G_f,h_{\pi(f)})]$$
$$= \sum_f \left(h_f - h_{\pi(f)}\right) D(G_f) + \sum_f [(h_f)_- - (h_{\pi(f)})_-] E_f$$

where $(h)_- = \min(0,h)$. Two summations by parts (Lemma 2.1) give

$$\sum_f h_f D(g_f) + \sum_f (h_f)_- e_f = \sum_f h_f D(g_f,h_f)$$

and (6.7c) follows from (6.7a). The implication (6.7b) → (6.7d) is similar and is the content of Theorem 2.2.

c) (6.7e) follows from (6.7d) and the equality $D(G_f) = \delta(G_f)$.

d) (6.8) follows from (6.7c) and

$$\sum_{f \in \mathfrak{F}} \Big[h_f D(G_f,h_f) - h_{\pi(f)} D(G_f,h_{\pi(f)}) - \big(h_f \delta(G_f,h_f) - h_{\pi(f)} \delta(G_f,h_{\pi(f)})\big)$$
$$+ (h_f - h_{\pi(f)}) \sum_{v \in \mathcal{V}(G_f)} \delta(v)\Big]$$

$$= \sum_f \left[(h_f)_- - \left(h_{\pi(f)}\right)_- \right] B_f \qquad \text{(by (6.5b))}$$

$$= \sum_f (h_f)_- \, b_f \qquad \text{(summation by parts)}$$

$$= \sum_{v \in \mathcal{V}_0^e(G)} h_{\pi(v)} (d - \Delta^e(v)) \qquad \text{(by (6.5c))} .$$

∎

As an example of these bounds consider the massless ϕ_4^4 model, the tree

$\tau =$ 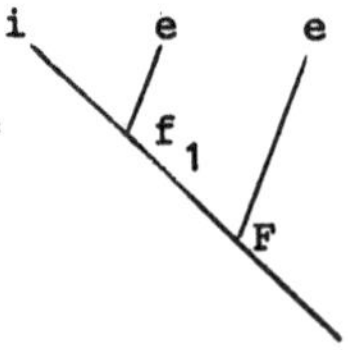

$$h_F < h_1 < 0 \tag{6.10a}$$

and the associated graphs (the vertices v_1 and v_2 in $\mathcal{V}_0^e(G_F)$ are marked with an x)

$G_F =$ 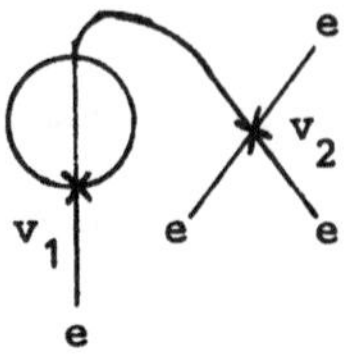

$G_{f_1} =$ 

Then

$$D(g_{f_1}, h_1) = D(G_{f_1}, h_1) = 2 \qquad \delta(G_{f_1}, h_1) = -1$$

$$D(g_F, h_F) = 2 \qquad D(G_F, h_F) = 4 \qquad \delta(G_F, h_F) = 0$$

and so the (equal) exponents in the bounds (6.7a), (6.7c) and (6.8) read

$$\sum h_f D(g_f, h_f) = 2h_1 + 2h_F$$

$$2(h_1 - h_F) + 4h_F \tag{6.10b}$$

$$-(h_1 - h_F) + 3h_1 + h_F .$$

The sum of the bound over $-\infty < h_F < h_1 < 0$ converges.

On the other hand for the tree

$\tau' =$ 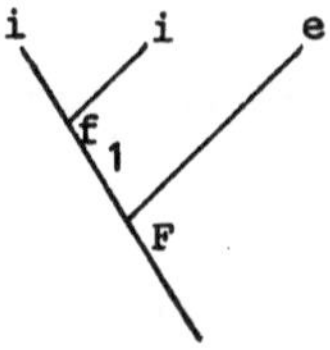

$$h_F < h_1 < 0$$

and the graphs

$G'_F =$ 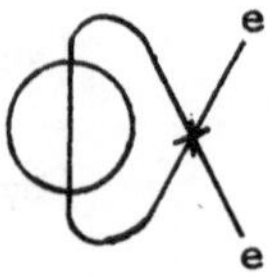

$G'_{f_1} =$ 

we have

$$D(g'_{f_1},h_1) = D(G'_{f_1},h_1) = \delta(G'_{f_1},h_1) = 2$$

$$D(g'_F,h_F) = 0 \qquad D(G'_F,h_F) = 2 \qquad \delta(G'_F,h_F) = 0 \ .$$

The equal exponents in (6.7a), (6.7c) and (6.8) are

$$2h_1, \quad 2(h_1-h_F) + 2h_F, \quad \text{and } 2(h_1-h_F) + 2h_F,$$

and the sum over h_F diverges.

There are several lessons to be drawn from these simple examples. The most important is that not much renormalization is needed in the IR regime! Even the divergent graph G'_F is only logarithmically divergent. We could choose to renormalize its subgraph G'_{f_1} $\left(\delta(G'_{f_1}) = 2\right)$ by a Taylor subtraction of order 2, but a Taylor subtraction of order 0 would do. For such a subtraction would introduce a factor $M^{-h_1+h_F}$, the bad factor M^{-h_1} coming from the Δx and the good factor M^{h_F} from the ∂ acting on a line of g'_F. This factor gives convergence:

$$\sum_{h_F<h_1<0} M^{2h_1} M^{-h_1+h_F} < \infty \ .$$

Moreover, external subgraphs don't need any renormalization. The <u>internal</u> subgraph G'_{f_1} needs renormalization because it can peel away with only (non-integrable) $|x|^{-4}$ decay from the rest of the graph, but the <u>external</u> subgraph G_{f_1} is pinned down by its external vertex.

In terms of scale decompositions, the sum over every h_f at an R-fork, i.e. over $h_{\pi(f)} < h_f < 0$, is finite. The only question is this: after all h_f's, $f > F$, have been summed, are we left with a positive coefficient for h_F when $h_F < 0$?

A bound like (6.8) is typically summed over scales h_f with $h_{\pi(f)} < h_f < \infty$. To do such sums, we note the following easily derived bound:

<u>Lemma 6.3.</u> Let $\alpha > 0$ be given. If a(h) depends only on the sign of h and satisfies $a(h \geq 0) \leq -\alpha$, then

$$\sum_{h_f > h_{\pi(f)}} M^{h_f a(h_f) - h_{\pi(f)} a(h_{\pi(f)})} (|h_f|+1)^{\kappa} \leq c \begin{cases} (|h_{\pi(f)}|+1)^{\kappa} & \text{if } a(h\leq 0) \leq -\alpha \\ (|h_{\pi(f)}|+1)^{\kappa+1} & \text{if } a(h<0) \leq 0 \end{cases} \tag{6.11}$$

where the number c depends on M, α and $\kappa \geq 0$.

The bounds of Theorem 6.2 parts c) and d) may be visualized by interpreting each factor of the bounds as a spring joining neighbouring forks of an extended tree τ^E of τ. For the bound (6.7e) when $h_F \geq 0$, τ^E is constructed by adding to τ a fork at its root whose scale is decreed to be zero.

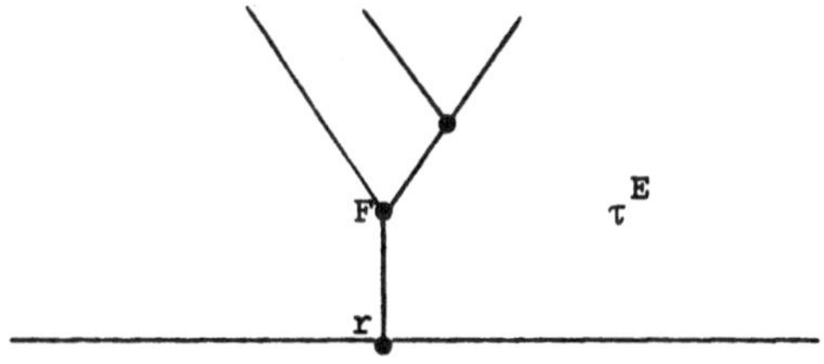

When the degree $\delta(G_f) < 0$ the spring factor $M^{(h_f - h_{\pi(f)})\delta(G_f)}$ ($M^{h_F \delta(G)}$ for f = F) "prevents the scales h_f and $h_{\pi(f)}$ of the forks at its ends from moving far apart". This is merely a colourful way of interpreting the bounds

$$\sum_{h_f = h_{\pi(f)}+1}^{\infty} M^{(h_f - h_{\pi(f)})\delta(G_f)} \leq 1 \qquad \text{if } \delta(G_f) < 0 \tag{6.12a}$$

$$\sum_{h_F = 0}^{\infty} M^{h_F \delta(G)} \leq 2 \qquad \text{if } \delta(G) < 0\,. \tag{6.12b}$$

(obtained from (6.11) with $\alpha = 1/2$ and $M \geq 4$).

For the bound (6.8) when $h_F < 0$, τ^E is constructed by adding to τ a new fork at each of its end points corresponding to vertices $v \in \mathcal{V}_0^e(G)$. These new forks are also decreed to have scale zero. For the example (6.10), τ^E is

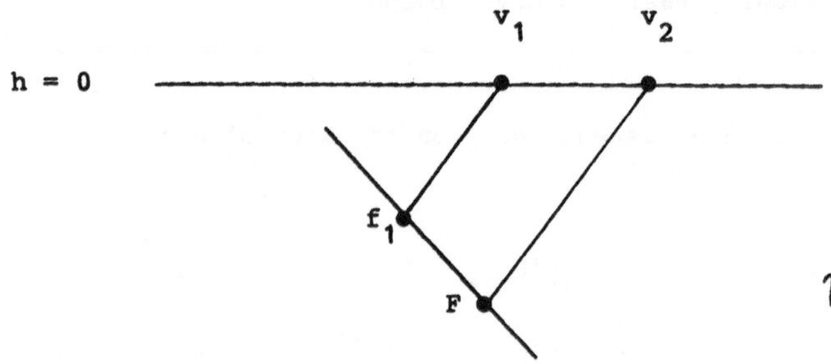

$$\mathcal{V}_0^e = \{v_1, v_2\}$$

Again, when the appropriate degrees are negative the spring factor $M^{h_f\delta(G_f,h_f)-h_{\pi(f)}\delta(G_f,h_{\pi(f)})}$ (or $M^{h_{\pi(v)}(d-\Delta^e(v))}$ for $f = v \in \mathcal{V}_0^e(G)$) keeps h_f and $h_{\pi(f)}$ close together.

The factors in the bound (6.8) for forks $f > F$ are more favourable than those in (6.7c), since $\delta(G_f,h_f) \leq D(G_f,h_f)$. Indeed, when $h_f < 0$ and $v^e(G_f) > 0$, we may have $D(G_f,h_f) \geq 0$. In this event the scale sum $\sum_{h_f=h_{\pi(f)}+1}^{\infty}$ is not uniform in $h_{\pi(f)}$ for (6.7c) while it is for (6.8): if $\delta(G_f) < 0$,

$$\sum_{h_f=h_{\pi(f)}+1}^{\infty} M^{h_f\delta(G_f,h_f)-h_{\pi(f)}\delta(G_f,h_{\pi(f)})} \leq 1 \tag{6.13}$$

whether $v^e(G_f)$ is positive or zero, provided $M \geq 4$. (We have used (6.11) with $\alpha = 1/2$ after noting that in dimensions $d \geq 3$ all degrees $\delta(G_f)$, $\Delta^i(G_f)$, etc. are half-integers).

As the price we must pay for making the coefficients $\delta(G_f,h_f)$ more negative for $f > F$, the first factor in the bound (6.8) fails to provide a factor for the bottom fork F since $\delta(G,h_F) = 0$. (See the example (6.10).) However, we can always recover some decay in h_F by choosing a chain of forks $f_1, \ldots, f_n \in \mathfrak{F}(\tau^E)$ with $\pi(f_1) = F$, $\pi(f_2) = f_1, \ldots, \pi(f_n) = f_{n-1}$ and $h_{f_n} \geq 0$. (If $h_f < 0$ for every $f \in \mathfrak{F}(\tau)$ then f_n will be an endpoint $v \in \mathcal{V}_0^e(G)$.) Suppose the degrees

$\delta(G_{f_i}, h_{f_i}) < 0$ for all the $f_i \in \mathfrak{F}(\tau)$. Then

$$\delta(G_{f_i}, h_{f_i}) \leq -\frac{1}{2} \text{ and } \delta\left(G_{f_i}, h_{\pi(f_i)}\right) \leq -\frac{1}{2} \tag{6.14}$$

and so we may steal a bit of the associated decay to construct a factor

$$\prod_{i=1}^{n} M^{-\frac{1}{4}\left(h_{f_i}-h_{\pi(f_i)}\right)} = M^{-\frac{1}{4}\left(h_{f_n}-h_F\right)} \leq M^{\frac{1}{4}h_F}$$

which renders the sum $\sum_{h_F=-\infty}^{-1}$ convergent without destroying the convergence of the sum (6.13). Thus for $h_F < 0$ and V dimensionless, we can use the bound

$$\|G\| \leq K^{\ell(G)} M^{\frac{1}{4}h_F} \prod_{i=1}^{n} M^{\frac{1}{4}\left(h_{f_i}-h_{\pi(f_i)}\right)} \prod_{f>F} M^{h_f\delta(G_f,h_f)-h_{\pi(f)}\delta\left(G_f,h_{\pi(f)}\right)}$$

$$\prod_{v\in\mathcal{V}_0^e(G)} M^{h_{\pi(v)}\Delta^i(v)}, \tag{6.15a}$$

and the extra factors $M^{\frac{1}{4}\left(h_{f_i}-h_{\pi(f_i)}\right)}$ do not affect the uniformity of the scale sums for f>F. In fact, we shall use the following simpler bound for $h_F < 0$:

$$\|G\| \leq K^{\ell(G)} M^{\frac{1}{4}h_F} \prod_{f>F} M^{h_f\delta(G_f,h_f) - h_{\pi(f)}\delta\left(G_f,h_{\pi(f)}\right)} M^{\frac{1}{4}\left(h_f-h_{\pi(f)}\right)}. \tag{6.15b}$$

The moral is that every sum $\sum_{h_f>k}$ is convergent <u>uniformly in k</u> if the conventional ultraviolet degree of divergence $\delta(G_f) < 0$ or if h_f is restricted to $h_f < 0$ when $v^e(G_f) \neq 0$.

Note that convergence is not ruined by subgraphs G_f with $\delta(G_f) = 0$ and $v^e(G_f) = 0$ provided h_f is restricted to $h_f < 0$. This is because while

$$\sum_{h_f=h_{\pi(f)}+1}^{-1} M^{\left(h_f-h_{\pi(f)}\right)\delta(G_f)} = |h_{\pi(f)}-1| \leq |h_F|$$

can generate powers of h, the powers of h will eventually be controlled by an exponential. Hence it is not necessary to renormalize marginal subgraphs in the infrared regime although we shall do so in order that (6.14) holds at each R-fork and we may steal some decay for F. However, it would be unwise to insert marginal counterterms directly in $\delta V^{I,U}$ because the effective coupling constant for such counterterms would behave like $\sum_{h_f=I}^{h_{\pi(f)}} M^{0h_f} \sim h_{\pi(f)}-I$ which diverges as $I \to -\infty$.

Instead, we obtain marginal counterterms by inserting

$$0 = L^0 G_f^0 - L^0 G_f^0$$

at forks with $h_f < 0$. Here we denote by L^δ the portion of the localization operator that extracts terms of degree δ:

$$(L^\delta W)(\Phi) = \frac{1}{(d-\delta)!} \partial_\zeta^{d-\delta} [\zeta^d W(\Phi^\zeta)]$$

(see (4.20)), and by L^+ the portion that extracts terms of strictly positive degree: $L^+ = \sum_{\delta>0} L^\delta$.

We generalize the definitions (2.51) and (2.52) of the R- and C-operations applied to a fork f, namely,

$$R = \chi(h_f > h_{\pi(f)}) (1 - L) ,$$

$$C = -\chi(h_f \leq h_{\pi(f)}) L ,$$

as follows: at $f > F$

$$R = \chi(h_f > h_{\pi(f)}) (1 - L) \tag{6.16a}$$

and

$$C = -\chi(h_f \leq h_{\pi(f)}) [L^+ + \chi(h_f \geq 0)L^0] + \chi(h_{\pi(f)} < h_f < 0)L^0$$

$$\equiv C_- + C_+ , \tag{6.16b}$$

whereas at F (r is the root scale)

$$R = \chi(h_F > r) [1 - L^+ - \chi(h_F \geq 0) L^0] \tag{6.16c}$$

and

$$C = C_- = -\chi(h_F \leq r) [L^+ + \chi(h_F \geq 0)L^0] . \tag{6.16d}$$

Our tree notation is as in §2 (see the paragraph of (2.53)) except that the R and C labels now have the meaning (6.16), each endpoint bears an e/i label, we write each C-fork as a sum of a C_-- and a C_+-fork so that the label ρ_f at a fork f takes one of the three values R, C_- or C_+ (the label C denotes the sum $C_- + C_+$), and the scales attached to the forks of a tree $\tau,\vec{\rho}$ run over the set

$$h_{\pi(f)} < h_f \le U \qquad \text{if } \rho_f = R \text{ or } C_+$$

$$I \le h_f \le h_{\pi(f)} \qquad \text{if } \rho_f = C_-,$$

which we denote by $\mathcal{H}(\tau,\vec{\rho})$.

It is easy to check that with these definitions we obtain the same renormalized tree expansion as in Theorem 2.3:

Theorem 6.4. Define the operations R and C by (6.16). Define

$$\delta V^{I,U} =$$ 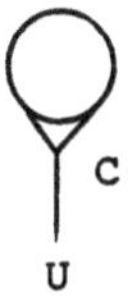 (6.17)

and

$$V_r^{I,U}(\phi^e) = \log \int \exp\left[(V+\delta V^{I,U})(\phi^{(r,U]}+\phi^e)\right] dP\left(\phi^{(r,U]}\right) + \text{const.} \tag{6.18}$$

Then

a) $$V_r^{I,U} =$$ 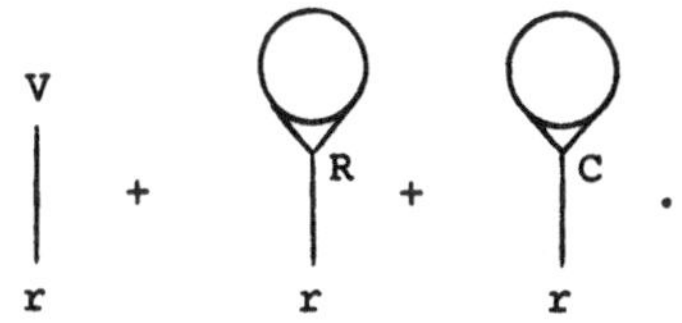 (6.19)

b)
$$\delta V_{n+1}^{I,U}(\phi^e) = -L_{n+1}^0 \log \int \exp\left[(V+\delta V_{\le n}^{I,U})(\phi^{[0,U]}+\phi^e)\right] dP\left(\phi^{[0,U]}\right)$$
$$-L_{n+1}^+ \log \int \exp\left[(V+\delta V_{\le n}^{I,U})(\phi^{[I,U]}+\phi^e)\right] dP\left(\phi^{[I,U]}\right) \tag{6.20}$$

for all $n \ge 1$ where the subscripts refer to the order of perturbation theory.

Remark. Suppose that every graph G_f with $\delta(G_f) > 0$ and $h_f < 0$ has at most 2 external legs. We call this property the "Biped Hypothesis". For example, massless ϕ_4^4 and QED_4 satisfy the Biped Hypothesis (but massless ϕ_3^6 does not). Then we may omit marginal counterterms in the IR regime (in other words apply (6.16c) and (6.16d) at all forks, not just F) and the proof of convergence of the resulting tree expansion is somewhat simpler than the proof of convergence of the more general tree expansion (6.19) that we give in Theorem 6.5 below. Omission of marginal counterterms in the IR regime in no way affects the effective potentials:

(6.18) and (6.20) still apply.

Proof.   a)   With the definition (6.16) we have at a fork $f > F$

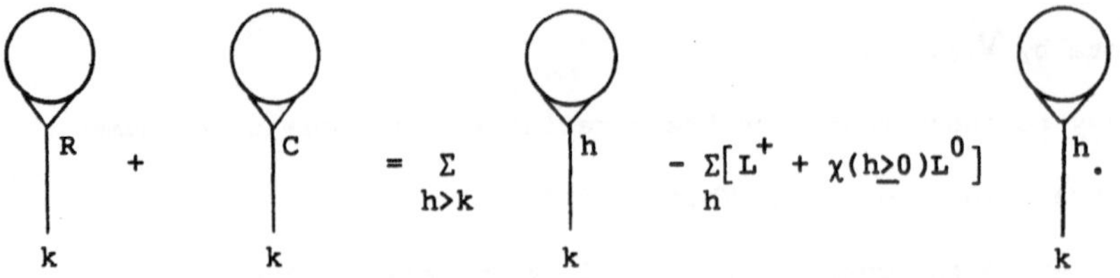

$$\text{[R diagram]}_k + \text{[C diagram]}_k = \sum_{h>k} \text{[h diagram]}_k - \sum_h \left[L^+ + \chi(h\geq 0)L^0 \right] \text{[h diagram]}_k. \qquad (6.21)$$

With (6.21) replacing (2.53), the proof of (6.19) is now identical to the proof of (2.56).

b)   Set $r = -1$ in (6.19) and apply L^0. The C-graphs on the right hand side drop out because of the $\chi(h_F \geq 0)$ in (6.16d) and so we obtain

$$L^0\, v_{-1}^{I,U}(\Phi^e) = L^0\, v(\Phi^e) . \qquad (6.22a)$$

Then set $r = e$ (in other words $I-1$) and apply L^+:

$$L^+\, v_e^{I,U}(\Phi^e) = L^+\, v(\Phi^e) . \qquad (6.22b)$$

As in Corollary 2.4, (6.20) follows from substituting (6.18) into (6.22) and projecting the resulting equations onto the order $n+1$ of perturbation theory.

▉

Note from (6.22) that the counterterms (6.17) are fixed by the following normalization conditions: Dimensionless parameters are fixed by a normalization condition at scale -1 (analogous to having external momenta of the order 1). Parameters with nonzero dimension (such as mass) are fixed by a normalization condition at scale $I-1 \to -\infty$ (i.e. at zero external momenta). In particular, if Φ_i is a massless field then its renormalized mass is also zero. If we wanted (or were even merely willing to end up with) a massive theory, it would make more sense to start off with a massive covariance, thereby eliminating the IR regime entirely. Finally, we remark that one cannot normalize dimensionless parameters at scale $-\infty$. Any graph contributing to such a normalization condition would be superficially IR divergent and consequently would have zero as an exceptional momentum.

We now turn to the question of the convergence of the tree expansion (6.19). For the remainder of this section we assume that the interaction V is dimensionless (see (2.84)) and that $d \geq 3$. Our strategy for bounding a tree τ contributing to (6.19) is based on the following considerations. At a fork $f \in \mathfrak{F}_p(\tau)$ (i.e. $h_f \geq 0$) our estimation procedure is the same as in Theorem 2.5. At a fork $f \in \mathfrak{F}_n$ (i.e. $h_f < 0$) we wish to exploit the improvement factors $M^{h_f e_f}$ of Theorem 6.2 in order to produce a positive coefficient for h_F when $h_F < 0$ (see the examples after Theorem 6.2). There, these factors were extracted at unrenormalized forks, but in fact they are not available at renormalized forks: at a C-fork we are unable to take advantage of the factor $M^{(h_f)_- e_f}$ because of the evaluation at zero momentum (equivalently, integration over all x_i's save one), and at an R-fork we likewise lose this factor because of the local counterterms implicit in the operation $R = 1 - L$.

The way out of this impasse is to remember that at an external, negative scale fork $f \in \mathfrak{F}_{en} \equiv \mathfrak{F}_e \cap \mathfrak{F}_n$, we don't actually need a renormalization cancellation. Accordingly, we decompose $R = 1 - L$ and estimate the 1 and $-L$ contributions separately, obtaining a factor $M^{(h_f)_- e_f}$ for the 1 contribution. Of course, if a fork $f \in \mathfrak{F}_{enR} \equiv \mathfrak{F}_{en} \cap \mathfrak{F}_R$ has a C-fork below it (or a -L-fork), then this factor is again unavailable and so there is no point in decomposing R. We call a fork f "trapped" if there is a $C_\pm$ or -L fork $f' \leq f$, and "untrapped" if not.

We can now outline our procedure for estimating τ:

1. We decompose R's starting from the bottom of the tree. If $F \in \mathfrak{F}_{enR}$ we decompose $R = 1 - L^+$, and otherwise we make no decomposition. Next we consider the forks f with $\pi(f) = F$. If $f \notin \mathfrak{F}_{en}$ or if f is trapped we perform no further decompositions at forks $f' \geq f$. If $f \in \mathfrak{F}_{en}$ and if f is untrapped, we decompose its $R = 1 - L$ and then continue decomposing up the tree in the same way (i.e., next at forks f' with $\pi(f') = f$). The result is to represent τ as a sum of at most $2^{|\mathfrak{F}_{enR}|} \leq 2^{|\mathfrak{F}(\tau)|}$ new trees τ' with the same structure and end point labels

as τ but with each fork f labelled $\rho_f = C_-, C_+, L'$ or R' where

$$L' = \begin{cases} -\chi(r<h_F<0)L^+ & \text{if } f = F \\ -\chi(h_{\pi(f)}<h_f<0)L & \text{if } f \in \mathfrak{F}_e,\ f \neq F \\ 0 & \text{if } f \in \mathfrak{F}_i \end{cases} \tag{6.23a}$$

and

$$R' = \begin{cases} 1 & \text{if } f \in \mathfrak{F}_{enu} \\ R & \text{if } f \notin \mathfrak{F}_{enu} \; . \end{cases} \tag{6.23b}$$

Here $\mathfrak{F}_{enu} = \mathfrak{F}_{en} \cap \mathfrak{F}_u$ where $\mathfrak{F}_u$ is the set of untrapped forks,

$$\mathfrak{F}_u(\tau') = \{f \in \mathfrak{F}(\tau') \,|\, \rho_{f'} = R' \text{ if } f' \leq f\} \; .$$

We denote the set of trapped forks $\mathfrak{F} \setminus \mathfrak{F}_u$ by $\mathfrak{F}_t$, and we shall use the symbol C' at a fork f to denote the sum over $\rho_f = C_-, C_+$ and L'.

In this way we obtain from (6.19) the modified tree expansion

$V_r^{I,U} =$ 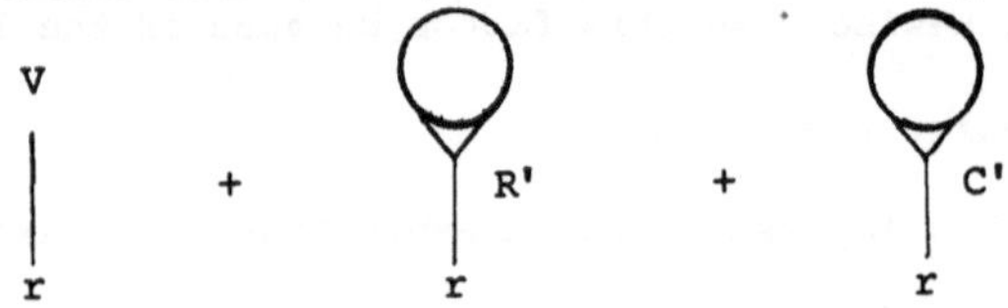

(6.24)

where the sums are over all trees τ (we drop the ') whose forks are labelled R' or C'. Of course, by construction, a fork f cannot be an L'-fork if there is a C'-fork below it.

The set of unrenormalized graphs $\mathcal{G}(\tau,\vec{\rho})$ associated with a C', R'-labelled tree τ is just as described before (2.75) except for the different meanings of the labels ρ_f. For instance, if $G_f^u = \int K(x)\,\Pi(x)\,dx$ is the subgraph of G at the fork f, not yet labelled by ρ_f, and if $\delta = \delta(G_f^u) \geq 0$ then

$$G_f = \begin{cases} G_f^u & \text{if } \rho_f = R' = 1 \\ \text{a monomial in (2.68)} & \text{if } \rho_f = R' = R \\ \text{a monomial in (2.71)} & \text{if } \rho_f = C_- \text{ or } L' \\ \text{a monomial in (2.71) with } m = \delta & \text{if } \rho_f = C_+ \; . \end{cases}$$

2. Given a graph $G \in \mathcal{G}(\tau,\vec{\rho})$ we bound $\|G_f\|$ at each minimal C'-fork f as in §2, i.e. by (2.86). (Here a minimal C'-fork is one with no C'-fork below it.)

This gives us a bound

$$\|G_f\| \le c_0^{\ell(G_f)} \prod_{f'>f} M^{\delta_{f'}(h_{f'}-h_{\pi(f')})} M^{\delta_f h_f} \tag{6.25}$$

where $\delta_f = \delta(G_f)$ includes the effect of Taylor operations.

3. We then adopt a "trimmed tree" point of view in which we regard each minimal C'-fork f_j in τ as contributing an interaction vertex $L^{\delta_j} G_{f_j}$ of dimension δ_j whose coefficient depends on the scales $h_{f'}$ $(f' \ge f_j)$ but is bounded as in (6.25). Of course, these interaction vertices may be dimensionful $(\delta_j > 0)$. Let $\tilde{\tau}$ denote the trimmed tree with these new vertices in place of the subgraphs G_{f_j}. For example:

$\tau =$ 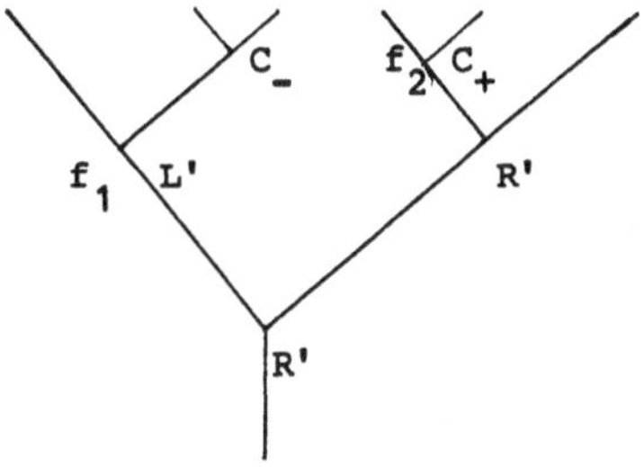 $\tilde{\tau} =$ 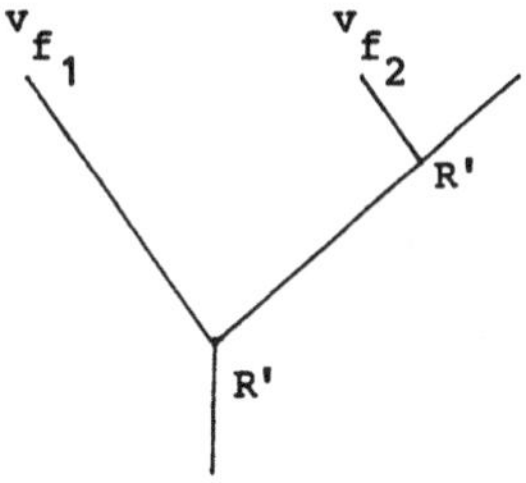 .

Note that $\tilde{\tau}$ has only R'-forks. The value of the original graph G is then given by the value of the corresponding graph $\tilde{G}$ associated with $\tilde{\tau}$ (whose vertices have coupling constants depending on the scales of the trimmed forks).

4. We next bound $\tilde{G}$ using (6.8) (and then (6.15b)). Of course, (6.8) was established for an unrenormalized graph, whereas $\tilde{G}$ comes with an R'-operation at each fork of $\tilde{\tau}$. Observe, however, that we do not attempt to extract a factor $M^{(h_f)_- e_f}$ and to make a renormalization cancellation at f simultaneously. According to the definition (6.23b) of R', if $f \in \mathcal{J}_{en}$ we extract the $M^{h_f e_f}$ as in Theorem 6.2 without introducing any new Taylor derivatives, whereas if $f \notin \mathcal{J}_{en}$ we make only a renormalization cancellation (now of course $M^{(h_f)_- e_f} = 1$).

If $f \in \mathcal{J}_u$, the bound (6.8) on $\|G_f\|$ contains the factor

$\prod_{v \in \mathcal{V}_0^e(\tilde{G}_f)} M^{h_{\pi(v)}(d-\Delta^e(v))}$. We shall make the identification

$$\mathcal{V}_0^e(\tilde{G}_f) \cong \mathcal{V}_0^{eu}(G_f) \cup \{f' \in \mathcal{J}_e \mid f'>f \text{ a minimal } C'\text{-fork of } \tau \text{ with } h_{\pi(f')}<0\}$$

and also set $\mathcal{V}_0^e(\tilde{G}_f) = \emptyset$ if $f \in \mathfrak{F}_t$.

With this notation in hand, we now establish the bound:

$$\|G_f\| \leq c_0^{\ell(G_f)} M^{X_f} \tag{6.26a}$$

where the exponent

$$X_f = \sum_{f'>f} [h_{f'} a_{f'}(h_{f'}) - h_{\pi(f')} a_{f'}(h_{\pi(f')})] + h_f a_f(h_f)$$
$$+ \sum_{v \in \mathcal{V}_0^e(\tilde{G}_f)} h_{\pi(v)}\left(d-\Delta^e(v)\right) \tag{6.26b}$$

with coefficients

$$a_f(h) = \begin{cases} \delta_f & \text{if } f \in \mathfrak{F}_t \cup \mathfrak{F}_i \quad \text{or} \quad h \geq 0 \\ -\Delta^i(G_f) & \text{if } f \in \mathfrak{F}_{eu} = \mathfrak{F}_e \cap \mathfrak{F}_u \quad \text{and } h < 0 \; . \end{cases} \tag{6.26c}$$

That is, we prove this extension of Theorem 2.5:

<u>Theorem 6.5.</u> Consider a Euclidean quantum field theory with dimensionless interaction in $d \geq 3$ dimensions, and let G be a graph associated with a tree τ in the renormalized tree expansion (6.24). Then its kernel is bounded by

$$\|G\| \leq c_0^{\ell(G)} M^{X_F} \tag{6.26d}$$

where c_0 is a constant independent of G, and X_f is defined in (6.26b).

<u>Proof.</u> The first step in the proof is to reread the proof of Theorem 2.5. We establish (6.26) in the same way as we did (2.83), proceeding by induction down the tree, bounding $\|K_f\|_\gamma$, where K_f is the kernel of G_f, in terms of the kernels $\tilde{K}_j$ of the graphs G_{f_j} associated with the forks $f_1, \ldots, f_p$ immediately above f. In this proof we regard the endpoints $v_1, v_2, \ldots$ of G_f as being distinct from these forks. Most

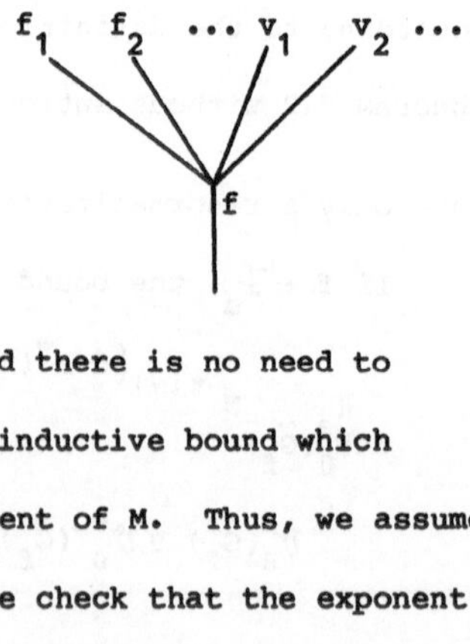

of the analysis is identical to that of Theorem 2.5, and there is no need to repeat it here. We need only examine that part of the inductive bound which differs from the bound of Theorem 2.5, namely the exponent of M. Thus, we assume that the exponent X_j at each f_j satisfies (6.26b) and we check that the exponent

X_f at f does too.

The induction actually begins with the highest untrapped forks since we have already seen in (6.25) that the inductive hypothesis holds for trapped forks. We apply the bound (2.99), modified by the replacement of $D(g_f)$ by $D(g_f,h_f)$ as in Theorem 6.2. The sum over $\vec{\alpha}$ is performed just as in Theorem 2.5 and we conclude that the exponent at f is bounded by (see 4) above)

$$\sum_{j=1}^{p} X_{f_j} + \bigl(D(g_f,h_f) - n_f\bigr)h_f \tag{6.27}$$

where n_f is the number of Taylor derivatives introduced at f $\bigl($the factor $(\Delta x)^{n_f}$ in K_f gives the contribution $-h_f n_f$ to (6.27)$\bigr)$. From the assumed form (6.26b) for X_{f_j} we find that $\sum_j X_{f_j}$ is not too different from the R.S. of (6.26b): we write

$$\sum_j h_{f_j} a_{f_j}(h_{f_j}) = \sum_j [h_{f_j} a_{f_j}(h_{f_j}) - h_f a_{f_j}(h_f)] + h_f \sum_j a_{f_j}(h_f)$$

and

$$\sum_j \sum_{v\in\mathcal{V}_0^e(\tilde{G}_{f_j})} h_{\pi(v)}\bigl(d-\Delta^e(v)\bigr) = \sum_{v\in\mathcal{V}_0^e(\tilde{G}_f)} h_{\pi(v)}\bigl(d-\Delta^e(v)\bigr)$$

$$- (h_f)_-\Bigl[\sum_{f_j\in\mathcal{J}_{eC'}} \bigl(d-\Delta^e(G_{f_j})\bigr) + \sum_{v\in\mathcal{V}_0^e(g_f)} \bigl(d-\Delta^e(v)\bigr)\Bigr] ,$$

and so find that

$$(6.27) = \text{R.S.}(6.26b) + b_f h_f$$

where

$$b_f = \sum_j a_{f_j}(h_f) - a_f(h_f) - \chi(h_f<0) \sum_{f_j\in\mathcal{J}_{eC'}} \bigl(d-\Delta^e(G_{f_j})\bigr) - \sum_{v\in\mathcal{V}_0^e(g_f)} \bigl(d-\Delta^e(v)\bigr) + D(g_f,h_f) - n_f . \tag{6.28}$$

We claim that $b_f = 0$ and so the desired bound (6.26b) on the exponent at f holds. To verify this claim first suppose that $f \in \mathcal{J}_i$ or $h_f \geq 0$. Then

$a_{f_j}(h_f) = \delta_{f_j}$, $a_f(h_f) = \delta_f$, $D(g_f,h_f) = D(g_f)$, and the third and fourth terms on the right of (6.28) are zero. Therefore

$$b_f = \sum \delta_{f_j} - \delta_f + D(g_f) - n_f = 0 \tag{6.29}$$

as in (6.6a). [The identity (6.6a) was derived for unrenormalized graphs; for renormalized graphs, δ_f includes the contribution $-n_f$ from the R-operation at f, and so (6.29) holds.]

Next suppose that $f \in \mathcal{F}_e$ and $h_f < 0$. Then

$$a_{f_j}(h_f) = \begin{cases} \delta_{f_j} & \text{if } f_j \in \mathcal{F}_i \cup \mathcal{F}_{C'} \\ -\Delta^i(G_{f_j}) = \delta_{f_j} + \Delta^e(G_{f_j}) - d & \text{if } f_j \in \mathcal{F}_{eR'} \end{cases}$$

$$a_f(h_f) = \delta_f + \Delta^e(G_f) - d ,$$

and

$$D(g_f, h_f) = D(g_f) + d\big(v^e(g_f) - 1\big) .$$

Therefore,

$$\begin{aligned} b_f &= \sum \delta_{f_j} + \sum_{f_j \in \mathcal{F}_e} \big(\Delta^e(G_{f_j}) - d\big) - \big(\delta_f + \Delta^e(G_f) - d\big) + \sum_{v \in \mathcal{V}^e(g_f)} \big(\Delta^e(v) - d\big) \\ &\qquad + D(g_f) + d\big(v^e(g_f) - 1\big) - n_f \\ &= 0 \end{aligned}$$

by (6.29) and the fact that

$$\Delta^e(G_f) = \sum \Delta^e(G_{f_j}) + \sum \Delta^e(v) .$$

∎

Note that when $h_F < 0$ and $\rho_F = C'$ the bound (6.26d) contains a spring for F since C' does not produce marginal counterterms according to the definitions (6.16d) and (6.23a) and so $a_F = \delta_F > 0$. As promised, we next produce a spring for F, when $h_F < 0$ and $\rho_F = R'$, by means of the inequality

$$\begin{aligned} \sum_{v \in \mathcal{V}_0^e(\tilde{G})} h_{\pi(v)}\big(d - \Delta^e(v)\big) &\le \sum_{v \in \mathcal{V}_0^e(\tilde{G})} h_{\pi(v)} \Delta^i(v) + \sum_{\substack{f > F \\ f \in \mathcal{F}_{L'}}} h_{\pi(f)} \delta_f \\ &\le \frac{1}{4} h_F + \frac{1}{4} \sum_{\substack{f > F \\ f \in \mathcal{F}_{eu}}} \big(h_f - h_{\pi(f)}\big) + \sum_{\substack{f > F \\ f \in \mathcal{F}_{L'}}} h_{\pi(f)} \delta_f \end{aligned} \tag{6.30}$$

where $\mathcal{F}_{L'}$ is the set of L'-forks of $\tilde{G}$ (or of G).

From (6.26) and (6.30) we obtain

$$\|G\| \le c_0^{\ell(G)}\, M^{\tilde{X}} \tag{6.31a}$$

where, for $h_F < 0$ and $\rho_F = R'$,

$$\tilde{X} = \sum_{f>F}\left[h_f a_f(h_f) - h_{\pi(f)} a_f(h_{\pi(f)})\right] + \frac{1}{4}\sum_{\substack{f>F\\ f\in \mathfrak{F}_{eu}}}\left(h_f - h_{\pi(f)}\right) + \frac{1}{4}h_F$$

$$+ \sum_{\substack{f>F\\ f\in \mathfrak{F}_{L'}}} h_{\pi(f)}\delta_f \tag{6.31b}$$

and, for $h_F \ge 0$ or $\rho_F = C'$,

$$\tilde{X} = \sum_{f>F}\left[h_f a_f(h_f) - h_{\pi(f)} a_f(h_{\pi(f)})\right] + \delta_F h_F\ . \tag{6.31c}$$

Our reason for setting aside the last term in (6.31b) is that when we sum over scales we shall need it in order to compensate for the wrong sign of the spring at an L'-fork.

The first two sums in (6.31b) or the sum in (6.31c) can be written (see the definition (6.26c) for a_f)

$$\sum_{\substack{f>F\\ f\notin \mathfrak{F}_{eu}}} \delta_f\left(h_f - h_{\pi(f)}\right) + \sum_{\substack{f>F\\ f\in \mathfrak{F}_{eu}}}\left[h_f a_f(h_f) - h_{\pi(f)} a_f(h_{\pi(f)}) + \frac{1}{4}\left(h_f - h_{\pi(f)}\right)\right]\ . \tag{6.32}$$

In the sum over $f \in \mathfrak{F}_{eu}$ (all R'-forks with $h_{\pi(f)} < h_f$)

$$[\ \dots\ ] = \begin{cases} (\delta_f + \frac{1}{4})(h_f - h_{\pi(f)}) & \text{if } h_{\pi(f)} \ge 0 \\ (\delta_f + \frac{1}{4})h_f + (\Delta_f^i - \frac{1}{4})h_{\pi(f)} & \text{if } h_{\pi(f)} < 0 \le h_f \\ (-\Delta_f^i + \frac{1}{4})(h_f - h_{\pi(f)}) & \text{if } h_f < 0 \end{cases}$$

where $\Delta_f^i = \Delta^i(G_f)$ and $\Delta_f^e = \Delta^e(G_f)$. Now

$$\delta_f = d - \Delta_f^e - \Delta_f^i \le d - \Delta_f^i \tag{6.33a}$$

and for $f \in \mathfrak{F}_{eu}$

$$\delta_f \le \frac{d+2}{2} - \Delta_f^i \tag{6.33b}$$

since $\Delta_f^e \ge \frac{d-2}{2}$.

It follows that at an R-fork where $\delta_f \le -1$,

$$\delta_f \leq - \frac{\Delta_f^i}{d+1} \tag{6.34a}$$

and that for $f \in \mathfrak{F}_{eu}$ and $h_f \geq 0$

$$\delta_f + \frac{1}{4} \leq - \frac{3}{2d+7} (\Delta_f^i - \frac{1}{4}) \ . \tag{6.34b}$$

In particular, the second sum in (6.32) is bounded by

$$- \frac{3}{2d+7} \sum_{\substack{f>F \\ f\in\mathfrak{F}_{eu}}} (\Delta_f^i - \frac{1}{4})(h_f - h_{\pi(f)}) \ ,$$

and the R-fork contribution to the first sum in (6.32) is bounded by

$$- \frac{1}{d+1} \sum_{\substack{f>F \\ f\in\mathfrak{F}_R \setminus \mathfrak{F}_{eu}}} \Delta_f^i (h_f - h_{\pi(f)}) \ .$$

Since $\Delta_f^i - \frac{1}{4} \geq \frac{2d-5}{2d-4} \Delta_f^i$ the contribution of all R'-forks to (6.32) is bounded by

$$- a \sum_{\substack{f>F \\ f\in\mathfrak{F}_{R'}}} \Delta_f^i (h_f - h_{\pi(f)})$$

where $a = \min\left(\frac{1}{d+1}, \frac{3(2d-5)}{(2d+7)(2d-4)}\right) > 0$ for $d \geq 3$.

In this way we obtain:

<u>Corollary 6.6.</u> If G is a graph associated with a tree τ in the renormalized tree expansion (6.24),

$$\|G\| \leq c_0^{\ell(G)} \prod_{f>F} M^{b_f(h_f - h_{\pi(f)})} M^{b_F h_F} \prod_{\substack{f>F \\ f\in\mathfrak{F}_{L'}}} M^{\delta_f h_{\pi(f)}} \tag{6.35}$$

where for $f > F$

$$b_f = \begin{cases} -a\Delta_f^i < 0 & \text{if } f \in \mathfrak{F}_{R'} \\ \delta_f \geq 0 & \text{if } f \in \mathfrak{F}_{C'} \end{cases} \tag{6.36a}$$

and

$$b_F = \begin{cases} \frac{1}{4} & \text{if } F \in \mathfrak{F}_{R'} \text{ and } h_F < 0 \\ \delta_F & \text{if } F \in \mathfrak{F}_{C'} \text{ or } h_F \geq 0 \ . \end{cases} \tag{6.36b}$$

Given the spring-loaded bound (6.35) it is now straightforward to sum over scales as in the proof of Theorem 2.6. We omit the details which are almost identical to those of §2 and which will be covered by our calculations in §8. As in §2, powers of $\vec{h}$ accumulate because of marginal C'-forks. Note that with our definition (6.16b) for $C_{\pm}$ we encounter the sum down to I:

$$\sum_{h_f=I}^{h_{\pi(f)}} M^{\delta_f(h_f-h_{\pi(f)})}(|h_f|+1)^{\kappa_f} \tag{6.37}$$

only for dimensionful counterterms ($\delta_f > 0$) and never for marginal counterterms ($\delta_f = 0$). We also observe that the contribution (6.37) to dimensionful counterterms (often just mass counterterms) from the infrared region is finite. But this finite piece is essential both for fixing the renormalized mass at zero and for convergence of the expansion under consideration.

We summarize the estimates that obtain in:

<u>Theorem 6.7.</u> Consider the renormalized tree expansion (6.24) for a Euclidean quantum field theory in d > 2 dimensions which has dimensionless interaction and involves massless fields. The contribution of each graph to (6.24) converges in the limits $U \to \infty$, $I \to -\infty$. Moreover:

a) A local term corresponding to a graph G associated with a renormalized tree τ in the sum

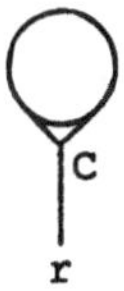

has the form $c\int P dx$ where P is a local Wick monomial in the fields and their derivatives with

$$\dim P \leq \begin{cases} d & \text{if } r \geq 0 \\ d-1 & \text{if } r < 0 \end{cases},$$

and the coefficient c satisfies

$$|c| \leq c_1^{\ell(G)} \kappa! \, (|r|+1)^{\kappa} M^{(d-\dim P)r} \tag{6.38}$$

where c_1 is independent of G and κ is the number of marginal C-forks in τ.

b) A graph G associated with a renormalized tree τ in the sum

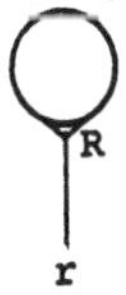

satisfies

$$\|G\| \leq c_1^{\ell(G)} \kappa!(r+1)^{\kappa} M^{\delta(G)r} \qquad \text{for } r \geq 0 \tag{6.39a}$$

where $\delta(G) < 0$, and

$$\lim_{r\to-\infty} \|G\| \leq c_1^{\ell(G)} \kappa! \; . \tag{6.39b}$$

As an example of a model subject to the above theorem we have:

<u>Corollary 6.8.</u> Consider the massless ϕ_4^4 field theory with propagator

$$C(x,y) = \frac{1}{(2\pi)^4} \int d^4k \, k^{-2} \, e^{ik\cdot(x-y)}$$

and interaction

$$V(\phi) = -\lambda \int :\phi^4(x): \, d^4x.$$

a) The contribution of any finite order in λ to the renormalized tree expansion (6.24) for the effective potential $V_r^{I,U}$ converges as $U \to \infty$, $I \to -\infty$ and $r \to -\infty$.

b) The contributions to 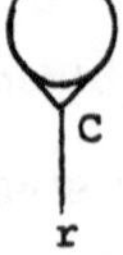

have the form

$$\begin{cases} \lambda_1(r) \int :\phi^2(x): \, dx + \lambda_2(r) \int :(\nabla\phi)^2(x): \, dx + \lambda_3(r) \int :\phi^4(x): \, dx & r \geq 0 \\ \lambda_1(r) \int :\phi^2(x): \, dx & r < 0 \end{cases}$$

where $\lambda_j(r)$ is given by a formal power series in λ, $\lambda_j(r) = \sum_{n=2}^{\infty} \lambda_j^{(n)}(r) \, \lambda^n$, whose coefficients satisfy

$$|\lambda_j^{(n)}(r)| \leq c_j^{(n)} (|r|+1)^{n-1} \, M^{2r} \qquad j=1$$

$$|\lambda_j^{(n)}(r)| \leq c_j^{(n)} (r+1)^{n-1} \qquad j = 2,3.$$

c) The contribution of a renormalized graph G to the tree expansion is bounded as in (6.39) where $\delta(G) = \min$ (4 - number of external legs - number of derivatives on these legs, -1).

§7. QED Without Cutoffs

In this section we use the results of §6 to show the existence of QED_4 in perturbation theory in the IR limit $I \to -\infty$[28], and to extend the results of §§3-5 (which applied only when $I = 0$) to show the gauge invariance of the renormalization.

Theorem 7.1 (a) The effective potentials $V_r(A,\psi,\bar{\psi}) = \lim_{I \to -\infty} \lim_{U \to \infty} V_r^{I,U}(A,\psi,\bar{\psi})$ and $V_e(A,\psi,\bar{\psi}) = \lim_{r \to -\infty} V_r(A,\psi,\bar{\psi})$ of QED_4, defined by the renormalized tree expansion of (6.24) with photon propagator

$$D_{\mu\nu}(x,y) = \delta_{\mu\nu}(2\pi)^{-4} \int k^{-2} e^{ik(x-y)} d^4k ,$$

electron propagator

$$S(x,y) = (2\pi)^{-4} \int (\not{k}+m)^{-1} e^{ik(x-y)} d^4k \qquad (m \geq 1),$$

and interaction,

$$V(A,\psi,\bar{\psi}) = -e \int :\bar{\psi}(x)\not{A}(x)\psi(x): d^4x,$$

exist in perturbation theory.

(b) The contribution from 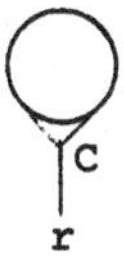

to V_r has the form $\sum_{j=1}^{7} \lambda_j(r)\, V_j(\Phi)$ (cf. (2.37)) where $\lambda_j(r)$ is given by a formal power series $\lambda_j(r) = \sum_{n=2}^{\infty} \lambda_j^{(n)}(r)\, e^n$ whose coefficients $\lambda_j^{(n)}(r)$ are bounded by

$c_j^{(n)}(|r|+1)^{n-1} M^{2r}$ for $-\infty < r < \infty$, $j = 6$ (photon mass)

$c_j^{(n)}(|r|+1)^{n-1} M^{r}$ for $-\infty < r < \infty$, $j = 1$ (electron mass)

$c_j^{(n)}(|r|+1)^{n-1}$ for $0 \leq r < \infty$, $j = 2,3,4,5,7,$

where the numbers $c_j^{(n)}$ are independent of r, e and m. The coefficients $\lambda_j(r)$,

$j = 2,3,4,5,7$, vanish for $r < 0$.

(c) The contribution of a renormalized graph G to V_r is bounded as in (6.39) with $\delta(G) = \min\big(4 - [\text{number of photon legs} + 3/2(\text{number of electron legs}) + \text{number of derivatives on these legs}], -1\big)$.

<u>Proof:</u> Theorem 6.7 applies. Note that we do not need the electron mass $m \geq 1$, i.e. the fact that $S^{(h)} = 0$ for $h < 0$: The power counting used in Theorem 6.5 requires only $|S^{(h)}(x,y)| < c\, M^{3h} \exp - M^h|x-y|$ for all h. The above bounds on V_r are thus uniform in the electron mass. ∎

We now argue that QED as defined in Theorem 7.1 is renormalized by gauge invariant counterterms provided $m \geq 1$: we use the results of §§3-5 extended to cases where the IR cutoff $I < 0$. The renormalized tree expansion for QED in the presence of the loop regularization Λ and cutoffs $I < 0$, U, N is given by (6.24) with the localization L replaced by the modified localization L^Λ of equation (3.31) (we write $L^\Lambda = L^0 + L^{+\Lambda}$). Note in particular the form of the order e^{n+1} counterterms (c.f. Theorem 6.4b):

$$\delta V^{I,U,\Lambda,N}_{n+1} = - L^0_{n+1}\, V^{I,U,\Lambda,N}_{-1,n} - L^{+\Lambda}_{n+1}\, V^{I,U,\Lambda,N}_{I-1,n} \tag{7.1}$$

where

$$V^{I,U,\Lambda,N}_{r,n}(\Phi^e) = \log \int \exp\, [V+\delta V^{I,U,\Lambda,N}_{\leq n}](\Phi^{(>r)}+\Phi^e)\, dP(\Phi^{(>r)})$$

for $r \geq I-1$. $\big($Note also that for $I-1 \leq r < 0$, $\Phi^{(>r)} = (A^{(r,U]}, \psi^{[0,N]}, \bar{\psi}^{[0,N]})$.$\big)$ Theorem 3.2 applies here with no essential change and shows the convergence of the $N \to \infty$ limits $V^{I,U,\Lambda}_k = \lim_{N\to\infty} V^{I,U,\Lambda,N}_k$ and $\delta V^{I,U,\Lambda} = \lim_{N\to\infty} \delta V^{I,U,\Lambda,N}$ as formal power series. From (7.1),

$$\delta V^{I,U,\Lambda}_{n+1} = - L^0_{n+1}\, V^{I,U,\Lambda}_{-1,n} - L^{+\Lambda}_{n+1}\, V^{I,U,\Lambda}_{I-1,n}\,. \tag{7.2}$$

Now the effective potentials renormalized up to order e^n, $V^{I,U,\Lambda}_{k,n}$, obey the Ward identity of Lemma 4.2, for all negative scales $I-1 \leq k \leq -1$. This is true provided $m \geq 1$, because the measure $d\nu_\Lambda\big(\psi^{(k,\infty)}\big)$ has the full spinor covariance

$S = (-i\partial\!\!\!/ \, 1 + M)^{-1}$ for all negative scales k. Therefore, with the identity (7.2), Theorem 4.1 holds for $I < 0$, saying that the counterterms $\delta V_{n+1}^{I,U,\Lambda}$ are again of gauge invariant form.

Finally, the arguments of §5 go through unchanged when $I < 0$. Theorem 5.3 gives the real field effective potentials $V_r^I(A, \psi_0, \bar{\psi}_0)$ as the limit of theories renormalized with gauge invariant counterterms. But the effective potentials V_r^I of Theorem 5.3 have the identical renormalized tree expansion as those of Theorem 7.1. We conclude:

<u>Theorem 7.2</u> For each $I < 0$ and in the limit $I \to -\infty$, QED_4 as defined in Theorem 7.1 is renormalizable using gauge invariant counterterms.

§8 Local Borel Summability

We prove in this section a bound on the large order behaviour of perturbation theory that is applicable to a broad class of models. It implies, in particular, that QED_4 is locally Borel summable.

We consider the class of all models having polynomial interactions each of whose vertices v is dimensionless, i.e. $\delta(v)=0$. Of course this renormalizability hypothesis may be weakened if some or all of the fields of the model have UV cutoffs or IR cutoffs (e.g. masses). For simplicity we assume that the interaction V is linear in the overall coupling constant λ and hence that terms of order n are those that contain n vertices (but see Remark 3 below). Note that the renormalized interaction $V+\delta V^{I,U}$ consists of V plus counterterms $\delta V^{I,U}$ that are always of order strictly greater than one.

Let $\lambda^n S_p^{(n)}$ be the contribution of order n in λ to the p-point Euclidean Green's function S_p (connected or not, amputated by the free propagator or not) or to the effective potential at any scale. In the sense of formal power series, $S_p \sim \sum_{n=0}^{\infty} \lambda^n S_p^{(n)}$. In the amputated case, we estimate $S_p^{(n)}$ (a sum of graphs) using the norm $\|\cdot\|$ defined in (2.21). In the unamputated case, we introduce free propagators into the norm (which we still denote by $\|\cdot\|$) to recover the amputated case. We prove (as in §6 we assume $d \geq 3$):

Theorem 8.1 Let Λ (>2) be the maximum number of fields in any vertex of the interaction V (e.g. $\Lambda = 3$ for QED_4 and $\Lambda = 4$ for ϕ_4^4). There exist constants K_p (independent of n), and R>0 (independent of p and n) such that

$$\|S_p^{(n)}\| \leq K_p R^{-n} (n!)^{\frac{\Lambda}{2}-1} . \tag{8.1}$$

Hence if $\Lambda \leq 4$ the Borel transform $\tilde{S}_p = \sum_{n=0}^{\infty} \frac{\lambda^n}{n!} S_p^{(n)}$ of S_p exists and is analytic at least in the disc $|\lambda| < R$, i.e. the model is locally Borel summable.

Remarks 1. For the ϕ_4^4 model, $\Lambda = 4$, and then (8.1) becomes $\|S_p^{(n)}\| \le K_p R^{-n} n!$. This bound is not surprising. Analogous results have been proven in both the UV[16] and IR[29] regimes.

2. For QED_4, $\Lambda = 3$, so that (8.1) becomes $\|S_p^{(n)}\| \le K_p R^{-n}(n!)^{1/2}$. With the cubic interaction $\lambda\bar{\psi}\not{A}\psi$ there are roughly $n!(n-1)(n-3)\ldots \equiv n!(n-1)!!$ labelled Feynman diagrams of order n and so $S_p^{(n)} \sim \frac{1}{n!}[n!(n-1)!!] \sim (n!)^{1/2}$. One might hope that there would be some additional cancellations from the fermionic nature of ψ. But in fact the Borel transform with respect to the fine structure constant $\alpha = \frac{\lambda^2}{4\pi}$ (which is a more natural coupling constant for QED than λ) should have a finite radius of convergence and in particular a "renormalon"[30,16] singularity on the positive real axis. The singularity is caused by the existence of individual graphs of order α^n that contain roughly n C-forks and consequently develop values of the order of n!. While renormalons would not necessarily prevent the existence of QED_4 they would prevent its (global) Borel summability.

3. We have, for convenience, assumed that all vertices in the interaction are of order one. There are models, like Yang-Mills$_4$, where this is not the case. In Yang-Mills$_4$ (with or without matter fields) the interaction contains three-legged vertices of first order and four-legged vertices of second order. In fact we prove a stronger bound than (8.1) which allows us to take advantage of the fact that some interaction vertices have order greater than one. This bound (8.13) says that if $S_p^{(m,L)}$ consists of those graphs having m vertices and L loops then

$$\|S_p^{(m,L)}\| \le K_p R^{-m} L! .$$

For Yang-Mills$_4$, graphs of order n have $L = \frac{n-p}{2} + 1$ and $m \le n$ so that

$$\|S_p^{(n)}\| \le K_p R^{-n} (n!)^{1/2}. \qquad (8.2)$$

This of course assumes that diagrams are renormalized according to their superficial degrees of divergence.

Proof of Theorem 8.1. It suffices to consider the p-point order n contribution to the effective potential at scale r and prove bounds of the type (8.1) uniformly in r. The tree expansion (6.24) expresses this contribution as a sum over all planar, unlabelled trees having a distinguished root, a single branch leaving the root, and n endpoints. Furthermore each fork f bifurcates into $p_f \geq 2$ lines. The number of such trees is bounded by 2^{4n}. (See, for example, Ref. 31, page 112.) Hence it suffices to consider a single tree τ.

We must also sum over all allowed R', C_+, C_-, L' labels on the forks of τ. There are at most 4^n such assignments so again it suffices to consider a single labelled τ. As in §2, each label generates a bounded number of monomials, c_1; altogether there are at most c_1^n terms generated and so it suffices to consider a single term. In particular we assume that each C' (i.e., C_+, C_-, or L') produces a specific counterterm.

In proving the bound (8.1) we find it convenient to adopt a "tree trimming" procedure similar to that used in Theorem 6.5. Recall that when a fork f is labelled C' (i.e. C_+, C_-, or L') the corresponding subgraph G_f is replaced by a counterterm, i.e. a vertex of interaction type multiplied by an effective coupling constant that depends on the scale $h_{\pi(f)}$. Thus a C'-fork has the same form as an endpoint. (We continue, as in Theorem 2.5, to include the endpoints of τ in the set of C'-forks.) We define the decomposition of τ into a <u>trimmed tree</u> $\tilde{\tau}$ and <u>counterterm insertion subtrees</u> $\sigma_1, \ldots, \sigma_m$ by cutting the branch beneath each minimal C'-fork (i.e. each C'-fork above F that is not above another C'-fork.) The resulting connected components are $\tilde{\tau}$ (the component which contains the root) and $\sigma_1, \ldots, \sigma_m$. We denote the minimal C'-forks by $f_1, \ldots, f_m$. For example, for

$\tau =$ 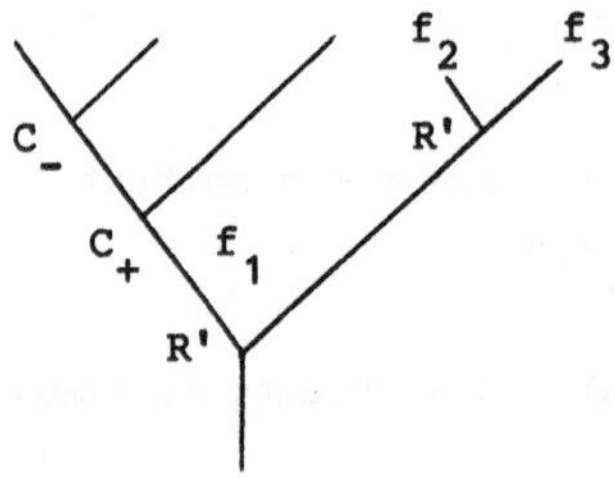

with degree n = 5, we have

$\tilde{\tau} =$ 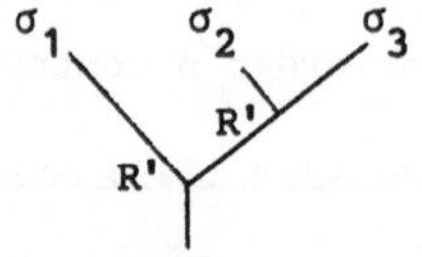

m=3

$\sigma_1 =$ 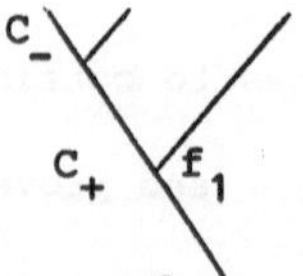

$n_1=3$

$\sigma_2 =$ 

$n_2=1$

$\sigma_3 =$ 

$n_3=1$.

Note that all the forks of $\tilde{\tau}$ are R'-forks. If $\tilde{\tau}$ has degree m and σ_i has degree n_i we have

$$n = \sum_{j=1}^{m} n_j .$$

The value of a labelled τ is given by

$$\sum_{\vec{h}(\tau)} \sum_{G} V^{\vec{h}(\tau)}(G) \tag{8.3a}$$

where $\vec{h}(\tau)$ is the scale assignment to the forks of τ (with root scale r), G runs over the graphs associated with τ, and $V^{\vec{h}(\tau)}(G)$ is the value of the graph G. With the above tree trimming procedure we can represent each G as a graph $\tilde{G}$ associated with $\tilde{\tau}$ which has m vertices, the jth vertex being the counterterm $L^{\delta_j} G_j$ from σ_j (here G_j is a graph associated with σ_j having external structure appropriate to the specific counterterm from σ_j). In this way, we can rewrite (8.3a) with self-evident notation as

$$\sum_{\vec{h}(\tau)} \sum_{\tilde{G}} \sum_{G_1,\ldots,G_m} V^{\vec{h}(\tau)}(\tilde{G};G_1,\ldots,G_m)$$

$$= \sum_{\vec{h}(\tilde{\tau})} \sum_{\tilde{G}} V^{\vec{h}(\tilde{\tau})}(\tilde{G};r(\sigma_1,h_1)W_1,\ldots,r(\sigma_m,h_m)W_m) \tag{8.3b}$$

where the last m arguments are the vertices of $\tilde{G}$:

$$r(\sigma_j,h_j)W_j = \sum_{\vec{h}(\sigma_j)} \sum_{G_j} V^{\vec{h}(\sigma_j)}(L^{\delta_j} G_j) . \tag{8.3c}$$

Here $\vec{h}(\sigma_j)$ is the scale assignment to the forks of σ_j with root scale $h_j = h_{\pi(f_j)}$, W_j is the integral of a local Wick monomial in $\Phi^{(\leq h_j)}$ (the jth vertex for $\tilde{G}$) and $r(\sigma_j,h_j)$ is the effective coupling constant that includes a sum over the scales $\vec{h}(\sigma_j)$ and over graphs G_j of order λ^{n_j}. Note that in our previous uses of an effective coupling constant at a C-fork on a trimmed tree, e.g.,

$\bar{U}_{f_j}(h_{\pi(f_j)})$ of (2.102) or the $L^{\delta_j}G_{f_j}$'s of Theorem 6.5, there was a single fixed graph G under consideration and so $\bar{U}_{f_j}$ or $L^{\delta_j}G_{f_j}$ did not involve a sum over subgraphs contributing to the counterterm insertion, as does $r(\sigma_j, h_j)$. The main burden of the present proof is to control this sum over graphs.

To control the sum over $\tilde{G}$ in (8.3b) we first estimate the number of unlabelled (i.e. topologically distinct) graphs contributing to $\tilde{\tau}$. Let there be m_j vertices with j legs. Then the total number of vertices is $m = \Sigma m_j$, of legs is $\ell = \Sigma jm_j$ and of loops is $L(\tilde{\tau}) = \frac{\ell-p}{2} - m + 1$. We note that $L(\tilde{\tau})$ is uniquely determined by the m_j's (i.e. by the vertices at the end points of $\tilde{\tau}$) and the fact that our graphs must be connected. The number of labelled graphs (i.e. with vertices labelled) is bounded by

$$\binom{\ell}{p} (\ell-p-1)!!$$

and the number of relabellings of the vertices is $\Pi\, m_j! \geq c_o^{-m}\, m!$. Since vertex-relabellings give topologically equivalent graphs the number of unlabelled graphs is bounded by

$$c_o^m\, (m!)^{-1} \binom{\ell}{p} (\ell-p-1)!! \leq c_2^m \left(\frac{\ell-p}{2} - m\right)_+! = c_2^m \left(L(\tilde{\tau})-1\right)_+!$$

where $(x)_+ = \max(0,x)$. Given an unlabelled graph G, there are a number of distinct labelled graphs obtained from G by assigning scales h_f and $<h_f$ to the hard and soft lines of G. As was shown in Appendix F of Reference 9, the number of ways of so labelling a given graph G compatibly with the tree $\tilde{\tau}$ such that the subgraph G_f associated with the fork f of $\tilde{\tau}$ has i_f internal legs is bounded by

$$n(\tilde{\tau})\, C_\varepsilon^m \exp\left[\varepsilon \sum_f i_f\right]$$

for any $\varepsilon > 0$, where $n(\tilde{\tau})$ was defined after (2.18a). Note that $i_f = |\Lambda^i(G_f)| \geq 1$ for $f > F$. The adjective "internal" does not occur in Appendix F of Reference 9 but, since there is never any question as to which legs are internal to G, we can put this adjective in for free.

We write the value (8.3b) of τ as

$$\sum_{\vec{h}(\tilde{\tau})} \sum_{\vec{i}} \sum_{\tilde{G}\in\mathcal{G}(\tilde{\tau},\vec{\rho},\vec{i})} V^{\vec{h}(\tilde{\tau})}(\tilde{G}; r(\sigma_1,h_1)W_1,\ldots) \tag{8.4}$$

where the second sum runs over $i_f \geq 1$ (for $f\in\mathcal{F}(\tilde{\tau})$, $f>F$), and the set $\mathcal{G}(\tilde{\tau},\vec{\rho},\vec{i})$ consists of graphs $\tilde{G}\in\mathcal{G}(\tilde{\tau},\vec{\rho})$ such that $\tilde{G}_f$ has i_f internal legs for $f>F$. By the estimates above, the number of such graphs is bounded by

$$|\mathcal{G}(\tilde{\tau},\vec{\rho},\vec{i})| \leq n(\tilde{\tau})\; c_2^m\; C_\varepsilon^m\; (L(\tilde{\tau})-1)_+!\; e^{\varepsilon\sum i_f}\;. \tag{8.5a}$$

Given a single graph $\tilde{G}$ we bound its contribution to (8.4) as in (6.35) but with the coupling constants $\prod_j r(\sigma_j,h_j)$ included and with the additional factor $\prod_{v\in\mathcal{V}(G)} M^{-h_{\pi(v)}\delta(v)}$ of (6.8) that arises from the fact that the vertices W_j may be dimensionful (i.e. $\delta_j = \delta(W_j) > 0$). The L' contribution to this factor cancels against the last factor in (6.35) to leave us with a net factor of $\prod_j M^{-h_j\eta_j}$ where

$$\eta_j = \begin{cases} \delta_j & \text{if } f_j \text{ is a } C_\pm\text{-fork} \\ 0 & \text{if } f_j \text{ is an L'-fork}\;. \end{cases}$$

Since all forks of $\tilde{\tau}$ are R'-forks and since $\Delta_f^i \geq \frac{d-2}{2}\, i_f$ we can bound the remaining exponent in (6.35) by

$$-\alpha|h_F| - 2\alpha\sum_{f>F} i_f(h_f-h_{\pi(f)}) \leq -\alpha|h_F| - \alpha\sum_{f>F}(i_f+h_f-h_{\pi(f)})$$

where $0 < \alpha \leq \frac{d-2}{4}\,a$. Hence altogether our bound on each labelled graph $\tilde{G}$ is

$$c_3^m\, M^{-\alpha|h_F|} \prod_{\substack{f\in\mathcal{F}(\tilde{\tau})\\ f>F}} M^{-\alpha(h_f-h_{\pi(f)})-\alpha i_f} \prod_{j=1}^m |r(\sigma_j,h_j)|\,M^{-h_j\eta_j}\;. \tag{8.5b}$$

Collecting together the bounds on the number of trees, R'/C₊/C₋/L' assignments, choices of terms generated and the bounds (8.5) on the number and individual contribution of graphs $\tilde{G}$, we have:

$$\|S_p^{(n)}\| \leq K_p 2^{4n} 4^n c_1^n \sup_{\tilde{\tau},\sigma_1,\ldots,\sigma_m} \frac{1}{n(\tilde{\tau})}\, c_2^m (L(\tilde{\tau})-1)_+!\; c_3^m \sum_{h_F} M^{-\alpha|h_F|}$$

$$\times \prod_{\substack{f\in\mathcal{F}(\tilde{\tau})\\ f>F}} \sum_{h_f>h_{\pi(f)}} M^{-\alpha(h_f-h_{\pi(f)})} \prod_{j=1}^m |r(\sigma_j,h_j)|\,M^{-h_j\eta_j} \sum_{i_f} n(\tilde{\tau})\; C_\varepsilon^m\; e^{\varepsilon\Sigma i_f} M^{-\alpha i_f}\,. \tag{8.6}$$

The constant K_p is 1 if $S_p^{(n)}$ is an effective potential and p! if $S_p^{(n)}$ is a Green's function. In the latter case it counts the number of labellings of the external lines of Feynman graphs. If we chose $\varepsilon < \alpha \ln M$ the sums over the i_f's converge so that

$$\|S_p^{(n)}\| \leq K_p\, c_5^n \sup_{\tilde{\tau},\sigma_1,\ldots,\sigma_m} c_4^m\, (L(\tilde{\tau})-1)_+!\, \sum_{h_F} M^{-\alpha|h_F|} \prod_{\substack{f\in\mathfrak{F}(\tilde{\tau})\\ f>F}} \sum_{h_f>h_{\pi(f)}} M^{-\alpha(h_f-h_{\pi(f)})} \tag{8.7}$$

$$\cdot \prod_{j=1}^{m} \frac{1}{n(\sigma_j)} |r(\sigma_j,h_j)|M^{-h_j\eta_j}$$

since the combinatoric factors satisfy $n(\tau) = n(\tilde{\tau}) \prod n(\sigma_j)$. We shall shortly prove by induction on the order n' of σ that

$$\frac{1}{n(\sigma)} M^{-h\eta}|r(\sigma,h)| \leq c_6^{n'-1} \quad \lambda_{\nu(\sigma)}(h) \tag{8.8}$$

where

$$\nu(\sigma) = \begin{cases} L(\sigma) & \text{for a dimensionless insertion} \\ L(\sigma)-1 & \text{for a dimensionful insertion} \end{cases} \tag{8.9}$$

and

$$\lambda_n(h) \equiv \sum_{i=1}^{\infty} (i+|h|+1)^n\, M^{-\frac{\alpha}{4}i} = \sum_{i>|h|} (i+1)^n\, M^{-\frac{\alpha}{4}(i-|h|)} . \tag{8.10}$$

(This λ_n is the same as the function defined in (2.108), except that the exponent of M is different.)

We can now show that $\nu(\sigma) \geq 0$. This can fail only in the case $L(\sigma) = 0$ (tree graphs). We may assume without loss of generality that no derivatives act on any line ℓ (otherwise move then onto external legs by repeated integration by parts: see equation (2.66).) Hence we always have $d_\ell < d$. Now any graph G with $L(G) = 0$ constructed purely from dimensionless vertices has $\delta(G) < 0$ and hence vanishing local part. This follows from (2.85) ($D = D^o$):

$$\begin{aligned}\delta(G) = D(G) - N_e &= \sum_{\ell\in\mathfrak{L}(G)} d_\ell - d(v(G)-1) - N^e \\ &= \sum_{\ell\in\mathfrak{L}(G)} (d_\ell - d) + dL(G) - N^e \\ &= \sum_{\ell\in\mathfrak{L}(G)} (d_\ell - d) < 0 .\end{aligned} \tag{8.11}$$

This ensures that $\nu(\sigma) \geq 0$.

$\lambda_n(h)$ satisfies the following properties. Properties a), c), d) and e) are also obeyed by the power function $(|h| + 1)^n$.

<u>Lemma 8.2</u> Fix any $\alpha > 0$.

a) $\prod_p \lambda_{n_p}(h) \leq \lambda_{\Sigma n_p}(h)$ if M is large enough.

b) $\sum_{h'>h} M^{-\frac{\alpha}{2}(h'-h)} \lambda_n(h') \leq \lambda_n(h)$ if M is large enough.

c) $\sum_{h=-\infty}^{\infty} M^{-\frac{\alpha}{2}|h|} \lambda_n(h) \leq c_7^n\, n!$

d) $\sum_{h'=h+1}^{-1} \lambda_n(h') \leq \lambda_{n+1}(h)$ for $h < -1$.

e) $\sum_{h'=0}^{h} \lambda_n(h') \leq \lambda_{n+1}(h)$ for $h \geq 0$.

f) $\sum_{h'\leq h} M^{-\frac{\alpha}{2}(h-h')} \lambda_n(h') \leq 2\,\lambda_n(h) \leq \lambda_{n+1}(h)$ if M is large enough.

g) $\lambda_n(h) \geq c_8^n\, n!$.

Before proving (8.8) and Lemma 8.2 let us show that they imply Theorem 8.1. Application of (8.8) to (8.7) gives

$$\|S_p^{(n)}\| \leq K_p\, c_5^n \sup_{\tilde{\tau},n_1,\ldots,n_m} c_4^m c_6^{n-m} (L(\tilde{\tau})-1)_+!\, \sum_{h_f} M^{-\alpha|h_F|} \prod_{f>F} M^{-\alpha(h_f-h_{\pi(f)})} \prod_{i=1}^{m} \lambda_{\nu(\sigma_i)}(h_i)$$

since $n = \sum_{i=1}^{m} n_i$. To perform the sum over the h_f's we apply Lemma 8.2a to combine the λ's at each maximal fork of $\tilde{\tau}$. We then apply Lemma 8.2b to "lower" the combined λ's to the next highest rank of forks of $\tilde{\tau}$. For example,

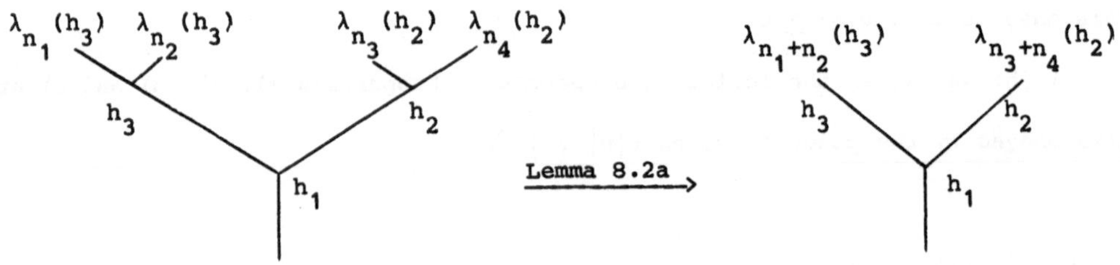

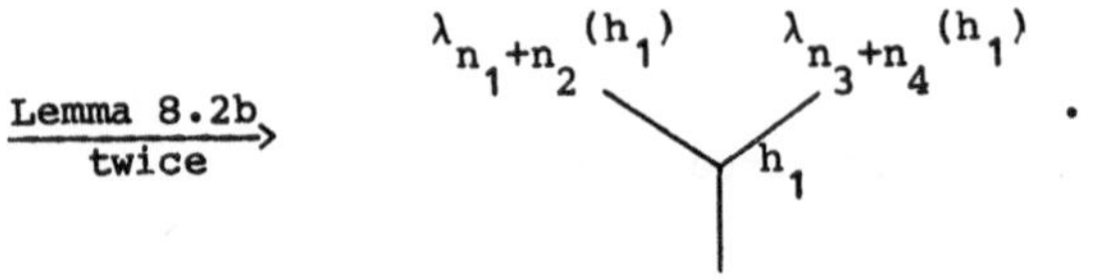

We repeat until all the λ's have been moved down to F:

$$\| S_p^{(n)} \| \le K_p \, c_5^n \sup_{n_i} c_4^m \, c_6^{n-m} \left(L(\tilde{\tau})-1\right)_+! \sum_{h_F} M^{-\alpha|h_F|} \lambda_{\Sigma\nu(\sigma_i)}(h_F) . \tag{8.12}$$

Then we apply Lemma 8.2c:

$$\begin{aligned} \| S_p^{(n)} \| &\le K_p \, c_5^n \sup_m c_4^m \, c_6^{n-m} \, c_7^{n-m} \, L(\tilde{\tau})! \left(\sum_{i=1}^m \nu(\sigma_i)\right)! \\ &\le K_p \, R^{-n} \left(L(\tilde{\tau}) + \sum_{i=1}^m L(\sigma_i)\right)! = K_p \, R^{-n} \, L(\tau)! \end{aligned} \tag{8.13}$$

with $R^{-1} = c_5 \max(c_4, c_6 c_7)$. Each vertex has at most Λ legs so $L(\tau) \le \frac{1}{2} \Lambda n - n + 1$ and (8.1) follows.

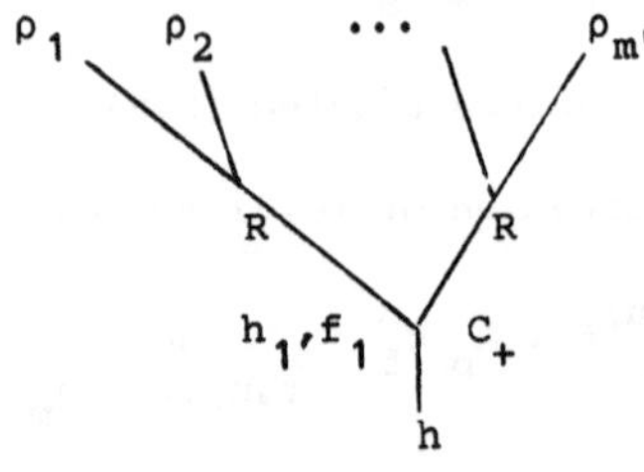

Example of a $\tilde{\sigma}$

Proof of (8.8). The bound on $r(\sigma,h)$ is developed in much the same way as the bound on τ we have just completed. σ is a tree of degree n' whose lowest fork f_1 is labelled C_+, C_- or L', whose forks $f>f_1$ are labelled C_+, C_- or R, and whose root scale is h. We decompose it into its trimmed part $\tilde{\sigma}$, of degree m', and "$C_\pm$ insertion subtrees" $\rho_1, \ldots, \rho_{m'}$ of degrees $n'_1, \ldots, n'_m$ by cutting the line below each minimal $C_\pm$ fork of σ. Note, however, that now the lowest fork f_1 of σ is necessarily a C' fork and is ignored when performing the cutting procedure. The argument leading up to (8.7) may now be repeated except that G_{f_1} is not renormalized, but instead a specific local part is taken from it. For a C'-fork with degree δ (see (6.16b) and (6.23a)) non-zero contributions come from h_1 summed over the range

$$\begin{cases} h+1 \text{ to } -1 & \text{if } h < 0 & (L') \\ 0 \text{ to } h & \text{if } h \geq 0 \text{ and } \delta = 0 & (C_-) \\ h+1 \text{ to } -1 & \text{if } h < 0 \text{ and } \delta = 0 & (C_+) \\ -\infty \text{ to } h & \text{if } \delta > 0 & (C_-) . \end{cases}$$

When we take a local part of G_{f_1}, the factor $M^{-\alpha|h_F|}$ in (8.6) is replaced by $M^{\delta h_{f_1}}$ (see (6.36b)). Hence the basic bound on $r(\sigma,h)$ is (c.f. (8.12)):

$$M^{-\eta h} |r(\sigma,h)| \leq n(\sigma) \sup_{n_i'} c_4^{m'} c_6^{n'-m'} \left(L(\tilde{\sigma})-1\right)_+! \sum_{h_1} M^{\delta h_1 - \eta h} \lambda_{\Sigma\nu(\rho_i)}(h_1). \tag{8.14}$$

Note that the number of trees, R, C_+, C_-, L' assignments and choices of counterterm types in σ have been counted in (8.6), and are not recounted here.

The sum over h_1 is bounded by use of various of the formulas of Lemma 8.2:

$$\sum_{h_1} M^{\delta h_1 - \eta h} \lambda_{\Sigma\nu(\rho_i)}(h_1) \leq \lambda_{\Sigma\nu(\rho_i)+\beta(\sigma)}(h) \tag{8.15}$$

where $\beta(\sigma) = \begin{cases} 0 & \text{if } \delta>0 \\ 1 & \text{if } \delta=0 \end{cases}$. For example, if f is a C_- fork with $\delta > 0$, we use part (b) and find

$$\sum_{h_1=-\infty}^{h} M^{\delta(h_1-h)} \lambda_n(h_1) \leq 2\lambda_n(h) .$$

Note that no 1 is gained in (8.15) when $\delta>0$ ("mass counterterms can't produce renormalons"). By Lemma 8.2g we have

$$\left(L(\tilde{\sigma})-1\right)_+! \leq c_9^{m'} \lambda_{(L(\tilde{\sigma})-1)_+}(h). \tag{8.16}$$

When (8.15) and (8.16) are inserted into the bound (8.14) and the λ's are combined using Lemma 8.2a we obtain

$$M^{-\delta h} \frac{|r(\sigma,h)|}{n(\sigma)} \leq \sup_{n_i'} 2\, c_4^{m'} c_6^{n'-m'} c_9^{m'} \lambda_{\tilde{\nu}(\sigma)}(h)$$

where $\tilde{\nu}(\sigma) = \left(L(\tilde{\sigma})-1\right)_+ + \Sigma\, \nu(\rho_i) + \beta(\sigma)$. If we choose $c_6 = \max(1, 2(c_4 c_9)^2)$, then (since $m' \geq 2$)

$$2\, c_4^{m'}\, c_6^{n'-m'}\, c_9^{m'} \leq c_6^{n'-m'/2} \leq c_6^{n'-1}.$$

To complete the proof of (8.8) it remains to verify that $\tilde{\nu}(\sigma) \leq \nu(\sigma)$. When $L(\tilde{\sigma}) \geq 1$ this is easy since then

$$\left(L(\tilde{\sigma})-1\right)_+ + \Sigma\, \nu(\rho_i) \leq L(\tilde{\sigma})-1 + \Sigma L(\rho_i) = L(\sigma) - 1.$$

When $L(\tilde{\sigma}) = 0$, i.e. the reduced Feynman graph g corresponding to $\tilde{\sigma}$ is a tree graph

$$\left(L(\tilde{\sigma})-1\right)_+ + \Sigma\, \nu(\rho_i) = L(\sigma) - \#\ \text{dimensionful vertices of } g.$$

But g must have at least one dimensionful vertex or else, by (8.11) its local part vanishes. Hence

$$\tilde{\nu}(\sigma) \leq L(\sigma) - 1 + \beta(\sigma) = \nu(\sigma).$$

■

Proof of Lemma 8.2

a) This is proven in Reference 32, Lemma 2.4f.

b) Since $\lambda_n(h)$ increases with $|h|$

$$\sum_{h'>h} M^{-\frac{\alpha}{2}(h'-h)} \lambda_n(h')$$

$$\leq \sum_{|h|\geq h'>h} M^{-\frac{\alpha}{2}(h'-h)} \lambda_n(h) + \sum_{h'>|h|} M^{-\frac{\alpha}{2}(h'-|h|)} \lambda_n(h')$$

$$\leq \left(M^{\frac{\alpha}{2}} - 1\right)^{-1} \lambda_n(h) + \sum_{\substack{h'>|h| \\ i>h'}} M^{-\frac{\alpha}{2}(h'-|h|)} (i+1)^n M^{-\frac{\alpha}{4}(i-h')}$$

$$\leq \frac{1}{2} \lambda_n(h) + \sum_{\substack{i>|h| \\ h'>|h|}} M^{-\frac{\alpha}{4}(h'-|h|)} (i+1)^n M^{-\frac{\alpha}{4}(i-|h|)}$$

$$= \frac{1}{2} \lambda_n(h) + \left(M^{\frac{\alpha}{4}} - 1\right)^{-1} \lambda_n(h)$$

$$\leq \lambda_n(h) \quad \text{if } M \text{ is large enough.}$$

c) $$\sum_{h=-\infty}^{\infty} M^{-\frac{\alpha}{2}|h|} \lambda_n(h) \le 2 \sum_{h=0}^{\infty} M^{-\frac{\alpha}{4} h} \lambda_n(h) = 2 \sum_{h=0}^{\infty} \sum_{i=1}^{\infty} (i+h+1)^n M^{-\frac{\alpha}{4}(i+h)}$$

$$= 2 \sum_{k=1}^{\infty} k(k+1)^n M^{-\frac{\alpha}{4} k} \le c_7^n \, n! \ .$$

d) $$\sum_{h'=h+1}^{-1} \lambda_n(h') \le |h| \, \lambda_n(h) \quad \text{(since } \lambda_n(h') \text{ increases with } |h'|\text{)}$$

$$\le \lambda_{n+1}(h) \ .$$

e) $$\sum_{h'=0}^{h} \lambda_n(h') \le (h+1) \, \lambda_n(h) \le \lambda_{n+1}(h) \ .$$

f) $$\sum_{h' \le h} M^{-\frac{\alpha}{2}(h-h')} \lambda_n(h') = \sum_{k=-h}^{\infty} M^{-\frac{\alpha}{2}(k+h)} \lambda_n(k) \quad \text{(where } k = -h'\text{)}$$

$$\le 2\lambda_n(h) \quad \text{(by part (b))}$$

$$\le \lambda_{n+1}(h) \quad \text{(since } (i+h+1) \ge 2 \text{ in (8.10))}.$$

g) $$\lambda_n(h) \ge \sum_{i=1}^{\infty} (i+1)^n M^{-\frac{\alpha}{4} i} \ge (n+1)^n M^{-\frac{1}{4}\alpha n} \ge c_8^n \, n! \ .$$

∎

It follows from Theorem 8.1 that

<u>Corollary 8.3</u> QED_4 is locally Borel summable.

Appendix A. Symbols and Terminology

Symbols

A

A	photon field	before (1.1)
A^{ζ}	scaled photon field	(4.22a)
$a_f(h)$	spring strength	(6.26c)
$a(U,M_0)$, a_n	spinor wave function counterterm coefficient	(3.29), (4.1)
α	fine structure constant	after (8.1)
α	parameter	before (8.5b)
α_f	$\|\{f' \in \mathfrak{F}_C \| f' > f, \delta_{f'} > 0\}\|$	after (2.108)
$\bar{\alpha}_f$	$\|\{f' \in \mathfrak{F}_C \| f' \geq f, \delta_{f'} > 0\}\|$	after (2.108)
α_n	$1 + a_{\leq n}$	after (4.5)

B

$B(\vec{x},\vec{h})$	bound on fermion loop	after (3.16)
B_f, b_f	relation between $D(G_f,h)$, $\delta(G_f,h)$	(6.5)
b	$\sum_{\ell \in \mathcal{L}(g_f)} d_\ell$	after (2.94)
b	parameter obeying $0 \leq b < [\max_f \Phi_f]^{-1}$	(5.21)
b_n	spinor mass counterterm coefficient	(3.29), (4.1)
β_f	$\|\{f' \in \mathfrak{F}_C \| f' > f, \delta_{f'} = 0\}\|$	after (2.108)
$\bar{\beta}_f$	$\|\{f' \in \mathfrak{F}_C \| f' \geq f, \delta_{f'} = 0\}\|$	after (2.108)
$\beta(\sigma)$	0 if $\delta > 0$, 1 if $\delta = 0$	(8.15)
$(x)_+$	$\max(0,x)$	before (2.87)
$(x)_-$	$\min(0,x)$	after (6.4b)

C

$\mathcal{C}$	test function space	(2.21b)

C_{ij}	$= \langle\Phi_i\Phi_j\rangle$, covariance or propagator	before (1.1)
$C_{ij}^{(h)}$	covariance of scale h	(1.3)
C_Λ	covariance	(3.12)
$C_m K$	m-counterterm part of K	(2.72)
 C	counterterm fork	(2.52),(6.16b)
$C_\pm$	counterterm fork	(6.16b)
C'	sum of C_+, C_-, L' labels	before (6.24)
c_n	charge counterterm coefficient	(3.29),(4.1)
c_1	maximum number of terms generated by R', $C_\pm$ or L'	after (8.2)
χ	gauge transformation	(1.18),(4.5)
χ^ζ	scaled χ	(4.22c)

D

D(x,y)	photon propagator	(1.11)
$D^{(h)}(x,y)$	photon propagator at scale h	(1.14)
D(G)	degree of divergence	(2.34),(2.36)
$D(G_{f,R})$	renormalized degree of divergence	before (2.44)
$D_s(G)$	spinor degree of divergence	before (3.15)
D(H,h)	degree of divergence in IR regime	(6.4a)
D^o	dimension not counting renormalization derivatives	(2.81)
$\mathcal{D}\Phi$	measure	(1.1)
$dP(\Phi)$	measure	(1.1),(1.12)
$dP_\Lambda(\Phi)$	free measure (loop regularized)	(3.12b)
$d\mu(A)$	free photon measure	before (1.11)
$d\nu(\psi,\bar\psi)$	free electron measure	after (1.11)
d_k	coefficient of F^2 counterterm	(4.2a)
d_ℓ	dimension of line	(2.22)

d^o	dimension not counting renormalization derivatives	(2.81)
$\det_2$	determinant	(3.1)
$\det_\Lambda$	loop regularized determinant	(3.9)
$\Delta^e(v), \Delta^e(H)$	dimension of legs external both to v,H and to G	(6.2b)
$\Delta^i(v), \Delta^i(H)$	dimension of legs external to v,H but internal to G	(6.2b)
$\Delta^{\delta+1}K(x)$	$\prod_{i=1}^{\delta+1} (x^{j_i}-x^0)K(x)$	(2.70)
$\tilde{\Delta}^\alpha(y)$	$\prod_{r=1}^{\alpha} (y^{i_r}-y^{j_r})$	(2.88)
δ_f	degree of counterterm	after (5.17)
δ^0	dimension not counting renormalization derivatives	(2.81)
δ_i	dimension of field	after (1.3)
	degree of counterterm	after (8.2)
δ_Λ	difference operator	(3.8)
$\delta(v)$	degree of vertex	(2.84)
$\delta(G)$	external degree of G	(2.45), (6.3b)
$\delta(H,h)$	external degree of divergence in IR regime	(6.4c)
$\delta\mathcal{L}^U$	counterterm part of Lagrangian	(2.37)
δV^U	counterterms	before (2.37)
δV^U_n	counterterms of order n	(2.60)
$\delta V^{I,U}$	counterterms	before (1.5)
$\delta V^{I,U}_{n+1}$	counterterms of order n+1	(6.20)
$\delta V_{\Lambda,\leq n}$	counterterms	(3.29),(3.33)
δV_n	counterterms	(4.1),(4.4)
$\delta V_{\leq n}$	counterterms	(4.3)
$\delta W_n(A)$	photon counterterms	(4.1),(4.2a)

E

$\mathcal{E}_h$	$\int \cdot dP(\phi^{(h)})$, expectation	(2.4)
$\mathcal{E}_{(k,h)}$	$\mathcal{E}_{k+1}\cdots\mathcal{E}_{h-1}$	before (2.11)
$\mathcal{E}_h^T$	truncated expectation	after (2.5)
E	SX	(4.7)
$\bar{E}$	XS	(4.7)
E_f	$d(v^e(G_f)-1)_+$	after (6.9)
E(H)	$d(v^e(H)-1)_+$	(6.4b)
e	electric charge	after (1.10)
e_f	$d(v^e(g_f)-1)_+$	after (6.9)
$\eta,\bar{\eta}$	spinor source	before (1.18)
η_i	δ_i for a C-insertion; 0 for an L'-insertion	before (8.5b)

F

$\mathfrak{F}(\tau)$	{forks of τ}	after (2.18a)
$\mathfrak{F}_R$	{R-forks}	(2.87)
$\mathfrak{F}_C$	{C-forks}	(2.87)
$\mathfrak{F}_{fR}$, $\mathfrak{F}_{fC}$, $\mathfrak{F}_{rR}$, $\mathfrak{F}_{rC}$	{fR or fC or fR or rC forks}	after (5.10)
$\mathfrak{F}_e, \mathfrak{F}_i$	{external forks}, {internal forks}	after (6.2)
$\mathfrak{F}_{eu}$	{external untrapped forks}	(6.26c)
$\mathfrak{F}_t$	{forks trapped by a C'-fork}	after (6.23)
$\mathfrak{F}_u$	{forks not trapped by a C'-fork	after (6.23)
F	lowest fork	Theorem 2.2
$F_{\mu\nu}$	electromagnetic field	(1.10)
ff	fictitious field	after (5.4)

	fork	(2.10)
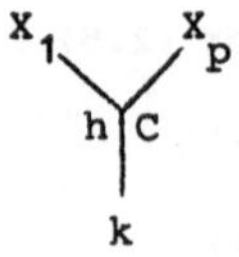	counterterm fork	(2.52), (6.16b)
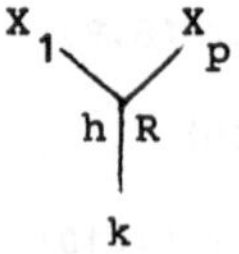	renormalized fork	(2.51), (6.16a)
$\Phi = \{\Phi_i\}$	fields	before (1.1)
$\Phi^{(h)}$	field of scale h	(1.2)
$\Phi^{(\leq U)}, \Phi^{(\geq I)}$, $\Phi^{[I,U]}, \Phi^{(k,U]}$	regularized fields	after (1.3)
Φ^{e}	external source	after (1.4)
$\Phi^{\zeta}(x)$	scaled field	(4.16)
Φ_f	$\phi(G_f/\{fR \text{ and } fC \text{ subgraphs of } G\})$	(5.14a)
$\phi(G)$	number of ff loops of G	(5.12)
ϕ_f	$\phi(g_f)$	(5.14b)

G

$\mathcal{G}(\tau)$	{graphs consistent with τ}	after (2.14)
$\mathcal{G}(\tau,\vec{\rho},G_0)$, $\mathcal{G}(\tau,\vec{\rho})$	{renormalized graphs}	before (2.75)
G	graph	(2.8a)
G_f^u	graph G_f with no R- or C-label at f	before (2.87)
$G_{f,R}$	renormalized value of graph	(2.39)
G^{Λ}	graph contributing to $V_r^{I,U,\Lambda}$	before (5.1)
$\tilde{G}$	graph G reduced by contracting subgraphs at minimal C'-forks	after (6.25)
$G(\Phi^e)$	generating functional of amputated connected Green's functions	(1.4)

g_f	reduced graph	after (2.14)
Γ	charge conjugation matrix	(3.39)
γ^μ	Euclidean Dirac matrices	after (1.10)

H

$\mathcal{H}(\tau,\vec{\rho})$	{scales}	(2.50)
$\tilde{\mathcal{H}}(\tau)$	{scales}	before (5.18)
H_f	product of propagators of g_f	(2.92)
h, $\vec{h}$, h_f	scales	(2.10),(2.13)

I

I	infrared cutoff	after (1.3)
i_f	$\lvert\Lambda(G_f)\setminus\Lambda(G)\rvert$	before (8.3a)

K

$K(q_1)$	fermion loop	Fig. (3.2)
κ	number of marginal C-forks in τ	Theorem 2.6
κ_f	$\lvert\{f'\in\mathfrak{F}'\,\vert\, f' > f\}\rvert$	after (2.89)
$\bar{\kappa}_f$	$\lvert\{f'\in\mathfrak{F}'\,\vert\, f' \geq f\}\rvert$	after (2.90)

L

$\mathcal{L}, \mathcal{L}_p, \mathcal{L}_f, \mathcal{L}_{int}$	Lagrangian densities	(1.10)
$\mathcal{L}_\Lambda$	Lagrangian density	(1.21),(3.11)
$\mathcal{L}(G)$	{lines of graph G}	after (2.89)
$\mathcal{L}_p(G)$	{photon lines of G}	before (5.12)
LG	local part of G	(2.40),(2.48)
$L_\delta = L^\delta$	localization operator of degree δ	(3.31), before (6.16)

L^+	$\sum_{\delta>0} L^\delta$	before (6.16)
$LV(\Phi)$	local part of V	(4.19)
L^Λ	loop regularized localization operator	(3.31)
L'	modified localization operator	(6.23a)
$L_f(x)$	length of tree $\mathcal{M}_f$	(2.76)
$L(\vec{x},\vec{h},m)$	fermion loop	before (3.16)
$L(q_1)$	fermion loop	Fig. (3.3)
L, $L(\tilde{\tau})$	$\lvert\{\text{loops}\}\rvert$	before (8.2)
ℓ	line of graph	after (2.8a)
$\ell(G)$	$\lvert\{\text{lines of } G\}\rvert$	before (2.19)
$\lvert\ell\rvert$	length of line ℓ	(2.23)
$\ell^{\vec{\mu}}(\vec{q},p,m)$	fermion loop	(3.3)
Λ	mass parameter for fictitious spinor fields	after (1.21), before (3.7)
Λ	maximum number of legs in interaction	before (8.1)
$\Lambda(G)$	{external legs of graph G}	after (2.8b)
$\Lambda(v)$	{legs of vertex v}	before (6.2)
$\Lambda^e(H)$, $\lambda^e(v)$	$\Lambda(H)\cap\Lambda(G)$, $\Lambda(v)\cap\Lambda(G)$	(6.2a)
$\Lambda^i(H)$, $\Lambda^i(v)$	$\Lambda(H)\setminus\Lambda(G)$, $\Lambda(v)\setminus\Lambda(G)$	(6.2a)
λ	gauge fixing parameter	after (1.10)
λ	external leg of graph	after (2.8b)
λ	coupling constant	Corollary 6.8,§8
$\lambda_n(h)$	$\sum_{i=1}^{\infty} (\lvert h\rvert+i+1)^n M^{-\alpha i/4}$	(2.108),(8.10)
λ_j	coupling constants	(2.37)
$\lambda_j(r)$	coupling constants	Theorem 7.1
$\lambda_j^{(n)}$	order n part of λ_j	after (2.37)
$\lambda_{5,n}$, $\lambda_{6,n}$, $\lambda_{7,n}$	counterterm coefficients	(4.14)

M

$\mathcal{M}$	minimal set of hard lines	(2.23)
$\mathcal{M}_f$	integration tree for g_f	before (6.9)
$M, M_{ij}=\delta_{ij}M_j$	mass matrix	after (3.30)
M	scale parameter	after (1.2)
M_i	mass of fictitious spinor field	(1.22), before (3.7)
m	electron mass	(1.10)
m	$\lvert\{$interaction/counterterm vertices$\}\rvert$	after (8.2)
m_i	mass of Φ_i	after (1.3)
m(s)	interpolating mass	before (3.17)
$m(\tau)$	combinatoric factor	(2.18b)
m_f	$\lvert\{$ff mass counterterms on external lines of G_f but not in any C-subgraphs$\}\rvert$	after (5.17)

N

N	UV cutoff for electron propagator	(1.20)
N_f	$\sum_{f' \geq f} n_{f'}$	before (2.81)
N_f^e	number of renormalization derivatives on external legs of G_f	(2.82)
n	degree of τ	after (8.2)
n_i	degree of σ_i	after (8.2)
$n(\tau)$	combinatoric factor	(2.18a)
n_f	degree of coordinate differences introduced by the Taylor operation at f	before (2.81)
$\lVert G\rVert$	norm	(2.21)
$\lVert G\rVert_0$	norm	after (2.76)
$\lVert K\rVert_\alpha$	norm	(2.88)

P

P_f	$\prod_{f'>f} M^{\left(D(G_{f'})-N_{f'}\right)\left(h_{f'}-h_{\pi(f')}\right)}$	after (2.89)

P_f	$p(G_f/\{\text{C-subgraphs of } G_f\})$	(5.13a)
$P_{a_1}(k_1)$	polynomial of degree a_1	(3.36)
p	p-point function	before (8.1)
p_f	number of branches growing up from f	before (2.15)
p_f	$p(g_f)$	(5.13b)
$p(G)$	$\sum_{\ell \in \mathcal{L}_p(G)} d_\ell$	(5.11)
$\Pi(x_1,\ldots,x_s)$	Wick monomial	before (2.48)
$\pi(f)$	fork below f	(2.19)
$\pi(v)$	fork below v	(6.5c)
$\Psi^e, \bar{\Psi}^e$	spinor source	after (4.5)
$\psi, \bar{\psi}$	electron fields	before (1.1)
$\psi_i, \bar{\psi}_i$	fictitious spinor fields	before (1.21), (3.10)
$\psi^\zeta, \bar{\psi}^\zeta$	scaled $\psi, \bar{\psi}$	(4.22b)

Q

Q	spinor mass and wave function insertion	before (3.34)
Q	$\bar{\psi}^e(\mathbf{XSX} + \mathbf{X})\psi^e$	(4.8)
$Q_j(\vec{p})$	polynomial of degree j	(3.4),(3.21)

R

R	radius of convergence of Borel transform	(8.1)
RG	renormalized graph	(2.49)
	renormalized fork	(2.51), (6.16a)
$R^\delta K$	renormalized kernel	(2.69)
R'	modified renormalization operation	(6.23b)
r	root scale	after (2.14)

rf	real field	before (5.3)
$r(\sigma_i,h_i)$	effective coupling constant	(8.3c)
ρ_f, $\vec{\rho}$	renormalization labels	(2.50), after (6.16), before (6.23)
ρ_i	$C_\pm$ insertion subtrees	after (8.13)

S

S	spinor propagator	after (4.1)
S(x,y)	electron propagator	before (1.1), after (1.11)
$S^{(h)}(x,y)$	electron propagator at scale h	(1.13)
S_p	p-point function	before (8.1)
$S_p^{(n)}$	n^{th} order contribution to p-point function	before (8.1)
$\tilde{S}_p$	Borel transform of S_p	after (8.1)
$S_p^{(m,L)}$	m-vertex, L-loop contribution to S_p	before (8.2)
$\sigma_1,\ldots,\sigma_m$	counterterm insertion subtrees	after (8.2)

T

T	factor in fermion loop	(3.19)
T_s	Taylor operator	(4.20)
T,t	transpose	after (3.39)
τ	tree	(2.13)
τ^E	extended tree	around (6.12)
$\tilde{\tau}$	trimmed tree	after (6.25), after (8.2)
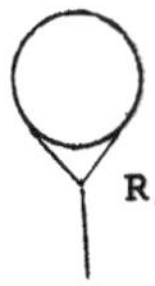 R,C or h r	sum of trees	after (2.51) before (6.17)

U

U	ultraviolet cutoff	after (1.3)
$U_f(h_f)$	$\sum_{\vec{h}} \prod_{f'>f} M^{\delta_{f'}(h_{f'}-h_{\pi(f')})}$	(2.103)
$\bar{U}_f(h_{\pi(f)})$	$\sum_{\vec{h}} \prod_{f'\geq f} M^{\delta_{f'}(h_{f'}-h_{\pi(f')})}$	(2.102)
$U_f(h_f)$	modified version of the above $U_f(h_f)$	(5.19)
$\bar{U}_f(k)$	modified version of the above $\bar{U}_f$	(5.20)

V

$\mathcal{V}(G)$	{vertices of G}	before (2.19)
$\mathcal{V}^e(H)$	{external vertices of H}	before (6.2)
$\mathcal{V}_0^e(H)$	$\{v\in\mathcal{V}^e(H)\mid h_{\pi(v)}<0\}$	before (6.8)
$\mathcal{V}_0^e(\tilde{G}_f)$	{external vertex or C'-vertex v of G_f that has no C'-fork of τ below it and has $h_{\pi(v)}<0$}	before (6.26a)
$\mathcal{V}_0^{eu}(G_f)$	$\{v\in\mathcal{V}_0^e(G_f)\mid$ there is no C'-fork of τ below v}	before (6.26a)
$V(\Phi)$	interaction potential	before (1.4), (1.12), (2.1)
$V_\Lambda(\Phi)$	loop regularized interaction potential	(3.12c)
$V_{e,un}$	unrenormalized effective potential	before (1.5)
V_e	renormalized effective potential	(1.5),(1.6)
$V_{e,n}$	order e^n contribution to effective potential	end of §1 (4.3)
$V_{o,n}$	effective potential in real field sector	after (4.13)
$V_e^{I,U}$	cutoff renormalized effective potential	(1.5),(1.6)
$V_k^{I,U}$, V_k^I	cutoff effective potential at scale k	(1.7)

$V_{k,\tau}^{I,U}$	contribution to $V_k^{I,U}$ from tree τ	(1.9)
V_k^U	cutoff effective potential at scale k	(2.2)
$V_{k,un}^{U,\Lambda,N}$	unrenormalized cutoff effective potential	(3.14)
$V_k^{U,\Lambda,N}$	cutoff effective potential at scale k	(3.33)
$V_{k,n}^{U,\Lambda,N}$	order e^n contribution to $V_k^{U,\Lambda,N}$	(3.33)
V_r^I, $V_r^{I,U}$	cutoff effective potential, $I=0$ $I\leq 0$	(5.2),(5.3) (6.18)
$V_r^{I,U,\Lambda}$	loop regularized, cutoff effective potential	Theorem 3.2
$V^s(\Phi)$	scaling dimension s part of V	(4.20)
V(G)	value of graph G	(2.8)
$V^{\vec{h}}(G)$	value of graph with given scales	before (2.17)
$V(\tau,\vec{h})$	value of tree	(2.17)
v(G)	$\lvert\{$vertices of G$\}\rvert$	before (2.19)
$v^e(G)$	$\lvert\{$external vertices of G$\}\rvert$	before (2.19)

W

$W(\eta,\bar{\eta},B)$	generating functional	before (1.18)
$W_n(A)$	photon counterterms	(3.29)

X

X	$e\not{\partial}\chi$	after (4.5)
X_f	exponent	(6.26b)
X_i	generalized vertices	before (2.9)
$x^j(t)$	interpolating position	after (2.67)

Z

Z	$\int e^{V(\Phi)} dP(\Phi)$, partition function	after (1.4)
Z	factor in fermion loop	(3.19)
ζ	scaling factor	(4.16)
$\zeta^h(k,m)$	factor in fermion loop	(3.19)

Terminology

Appendix B. Real Time

In the text we have dealt exclusively with the Euclidean (i.e. imaginary time) effective potentials and Green's functions. However, the object of direct physical significance in a quantum field theory is the relativistic or real time scattering matrix. In this appendix we explain how renormalization is carried out in real time. Our discussion is organized as follows:

1. We review the axiomatic construction of S-matrix elements from the renormalized Euclidean Green's functions (EGF's). Once the EGF's are shown to satisfy appropriate Euclidean axioms, then the Osterwalder-Schrader (OS) Reconstruction Theorem guarantees that they have an analytic continuation back to real time vacuum expectation values (VEV's). The S-matrix may then be obtained from these VEV's by the LSZ reduction formula.

2. Inasmuch as our renormalization of EGF's applies only to the formal power series of perturbation theory, we cannot return to real time courtesy of the OS Reconstruction Theorem. However, we can formally analytically continue each graph occurring in perturbation theory from the Euclidean to the relativistic world. This continuation, known as a "Wick rotation", is most conveniently done in α-p-space. Accordingly, we first re-express our renormalized graphs as integrals over p and Feynman parameters α. We have already made use of α-parameter integrals in the text but in a suppressed way (see Lemma 3.1). In terms of α-integrals, the decomposition of the (Euclidean) propagator $(-\Delta+m_\ell^2)^{-1}$ corresponding to the line ℓ may be written as

$$(-\Delta+m_\ell^2)^{-1} = \int_0^\infty d\alpha_\ell e^{-\alpha_\ell(-\Delta+m_\ell^2)} = \sum_{h=-\infty}^{\infty} \int d\alpha_\ell \chi_h(\alpha_\ell) e^{-\alpha_\ell(-\Delta+m_\ell^2)} \tag{B.1}$$

where $\{\chi_h\}$ is a suitable partition of $[0,\infty)$. In the text, for $m = 0$, we used

$$\chi_h = \text{characteristic function of } [M^{-2h}, M^{-2h+2}] \ . \tag{B.2a}$$

For $m > 0$ we used (B.2a) for $h > 0$, $\chi_h = 0$ for $h < 0$, and for $h = 0$,

$$\chi_0 = \text{characteristic function of } [1,\infty) \ . \tag{B.2b}$$

To obtain an α-parametric expression for the kernel of a graph, we simply insert (B.1) for every line (observing the scale orderings in force).

3. We then transfer the bounds of Theorem 2.5 on the Euclidean x-space kernels to bounds on the Euclidean α-x-space kernels. The argument, given in Corollary B.4, actually provides an alternate and simpler proof of Theorem 2.5. Of course, bounds do not analytically continue to relativistic space. Fortunately, though, the kernels of the Euclidean and relativistic graphs have factors in common to which our Euclidean bounds apply. In this way our control of relativistic p-space kernels is accomplished largely through Euclidean x-space bounds!

4. In general, convergence arguments are more delicate in the relativistic than in the Euclidean world because decay factors are replaced by oscillatory factors. For instance, in (B.1) with $m_\ell > 0$, $e^{-\alpha_\ell m_\ell^2} \rightarrow e^{-i\alpha_\ell m_\ell^2}$. Accordingly, to control the α-integrals for large α_ℓ in the relativistic setting we argue that the exponential decay for large α_ℓ in the Euclidean setting was not necessary for convergence and that a decay of $\alpha_\ell^{-2-\delta}$ would have sufficed. Such factors can be extracted in the relativistic setting from $e^{-i\alpha_\ell m_\ell^2}$ by integration by parts with respect to α_ℓ for $\alpha_\ell \geq 1$.

Thus our bounds on the renormalized graphs in the Euclidean world carry over to essentially the same bounds on real time graphs (our estimates on the graph-dependent constants are, however, not best possible in the relativistic world--see Theorem B.9). Throughout this appendix, for expository simplicity, we consider the case of a theory with a single neutral scalar field. All results extend easily to theories with more general fields.

The connection between the S-matrix and the real time VEV's is given by the LSZ reduction formula (Ref. 13, eqn. (5-28) or (5.50)). It gives the amplitude for scattering from a state consisting of ℓ particles of momenta $q_1,\ldots,q_\ell$ to a state consisting of n particles of momenta $p_1,\ldots,p_n$ as

$$\begin{aligned}&\langle p_1,\ldots,p_n \text{ out} \mid q_1,\ldots,q_\ell \text{ in}\rangle \\ &= (iZ^{-1/2})^{n+\ell} \int d^4y_1\ldots d^4x_\ell \exp\{i\sum_{k=1}^{n}\langle p_k,y_k\rangle - i\sum_{r=1}^{\ell}\langle q_r,x_r\rangle\} \\ &\quad \times (\Box_{y_1}+m^2)\ldots(\Box_{x_\ell}+m^2)(\Omega,\ T[\phi(y_1)\ \ldots\ \phi(x_\ell)]\Omega) \qquad \text{(B.3)} \\ &\quad + \text{disconnected terms}\ .\end{aligned}$$

Here everything is physical and in real time: ϕ is the physical real-time field, Ω is the physical vacuum, m is the physical mass, $\langle p,y\rangle$ and $\langle q,x\rangle$ refer to the Minkowski inner product and $\Box = \partial_0^2 - \vec{\nabla}^2$. The time-ordering operator T simply reorders the product of fields on which it acts so that their time arguments decrease from left to right. The constant Z is the "proportion of the two point function coming from the one particle mass shell". To be more precise it can be determined using the Källen-Lehmann representation (Ref. 13, eqn. (5-18))

$$(\Omega, [\phi(x),\phi(y)]\Omega) = iZ\Delta(x-y;m) + i\int_{m_1}^{\infty} d\rho(m')\ \Delta(x-y;m') \tag{B.4}$$

where $i\Delta(x-y;m)$ is the commutator for a free field of mass m. Finally there are simple explicit algebraic formulae relating connected and disconnected terms (Ref. 13, eqns (5-40) and (5-52)) that we won't worry about.

On the other hand the EGF's S_n are, by definition, the analytic continuation of the Wightman distributions

$$W_n(x_1,\dots,x_n) = (\Omega,\ \phi(x_1)\dots\phi(x_n)\Omega) \tag{B.5}$$

to imaginary time:

$$S_n(x_1,\dots,x_n) = W_n((ix_1^0,\vec{x}_1),\ \dots,\ (ix_n^0,\vec{x}_n)) \ . \tag{B.6}$$

(See Ref. 33, pp. 75-76.) The crucial ingredient for the existence of this analytic continuation is the fact that the joint spectrum of the energy-momentum operators $P = (H,\vec{P})$ lies in the closed forward light cone $\bar{V}_+$. This allows the analytic continuation of

$$W_n(x_1,\dots,x_n) = (\Omega,\ \phi(0)e^{i\langle x_2-x_1,P\rangle}\phi(0)\dots e^{i\langle x_n-x_{n-1},P\rangle}\phi(0)\Omega) \tag{B.7}$$

to

$$\{(x_1,\dots,x_n)\ |\ \mathrm{Im}(x_j-x_{j-1}) \in V_+\ ,\quad 2 \le j \le n\} \ .$$

Then Lorentz invariance, the existence of complex Lorentz transformations and locality may be used to extend the domain of analyticity to include the Euclidean points.

Given the family of (alleged) EGF's S_n of an (alleged) model one may use the Osterwalder-Schrader Reconstruction Theorem [33,34] to analytically continue back to the W_n's. Suppose that the S_n's satisfy the following conditions:

(E0) Regularity. It suffices that there exist constants α, β and a Schwartz space norm $|\cdot|$ such that

$$|S_n(f_1,\ldots,f_n)| \le \alpha(n!)^\beta \, \Pi \, |f_i| \;;$$

(E1) Euclidean invariance. For all $a \in \mathbf{R}_4$, $R \in SO_4$

$$S_n(\{x_i\}) = S_n(\{Rx_i + a\}) \;;$$

(E2) Reflection (OS) Positivity. For all finite sequences $f_0,\ldots,f_N$ of test functions with $f_0 \in \mathbf{C}$ and for $n \ge 1$

$$f_n \in \mathcal{S}(\mathbf{R}_<^{4n}) = \mathcal{S}(\mathbf{R}^{4n}) \cap \{f \mid \operatorname{supp} f \subset \dot{\mathbf{R}}_<^{4n}\}$$

where $\mathbf{R}_<^{4n} = \{\underline{x} \mid 0 < x_1^0 < x_2^0 \ldots < x_n^0\}$,

$$\sum_{n,m} S_{n+m}(\Theta f_n^* \times f_n) \ge 0$$

where $\Theta f_n(x_1,\ldots,x_n) = f_n((-x_1^0,\vec{x}_1),\ldots,(-x_n^0,\vec{x}_n))$;

(E3) Symmetry. For all permutations π,

$$S_n(x_1,\ldots,x_n) = S_n(x_{\pi(1)},\ldots,x_{\pi(n)}) .$$

Then the OS Reconstruction Theorem says that there exist corresponding W_n's related to the S_n's by (B.6) which obey:

(W0) Regularity. $W_n \in \mathcal{S}'(\mathbf{R}^{4n})$;

(W1) Relativistic invariance. For all $(a,\Lambda) \in P_+^\uparrow$

$$W_n(\{x_i\}) = W_n(\{\Lambda x_i + a\}) \;;$$

(W2) Positivity. For all finite sequences $f_0, f_1,\ldots,f_N$ of test functions $f_0 \in \mathbf{C}$, $f_n \in \mathcal{S}(\mathbf{R}^{4n})$, $n \ge 1$

$$\sum_{n,m} W_{n+m}(f_n^* \times f_m) \ge 0$$

where $f_n^*(x_1,\ldots,x_n) = \bar{f}_n(x_n,\ldots,x_1)$;

(W3) Locality. For any n and $k = 1,\ldots,n-1$

$$W_n(x_1,\ldots,x_k,x_{k+1},\ldots,x_n) = W_n(x_1,\ldots,x_{k+1},x_k,\ldots,x_n)$$

if $\langle x_k-x_{k+1},\ x_k-x_{k+1}\rangle < 0$.

The proof starts with a GNS-style construction in which (E2) is used to put a semidefinite inner product on a space of formal symbols. Factoring out null vectors and taking the closure gives the physical Hilbert space.

Once the Wightman distributions (B.5) have been constructed they have to be time ordered. Since, for example,

$$(\Omega,T[\phi(y_1)\phi(y_2)]\Omega) = \Theta(y_1^0 \le y_2^0)(\Omega,\phi(y_1)\phi(y_2)\Omega) + \Theta(y_1^0 > y_2^0)(\Omega,\phi(y_2)\phi(y_1)\Omega) ,$$

this amounts to showing that the W_n's are sufficiently regular as to allow the characteristic functions Θ in their test functions. Eckmann, Epstein and Fröhlich[35] show that this is indeed the case if one extends the OS positivity axiom (E2) to allow test functions

$$f_n \in \mathcal{S}(\mathbb{R}^{4n}) \cap \{f \mid \mathrm{supp}\ f \subset \{(x_1,\ldots,x_n) \mid x_i^0 \ge 0\}\} .$$

In a nonperturbative setting one would now have to verify that one indeed has an isolated mass hyperboloid with the desired mass m in the energy momentum spectrum, that one has asymptotic completeness in the low energy region, and that the distribution on the R.S. of (B.3) is sufficiently regular to allow the restriction of $p_1,\ldots,q_\ell$ to the mass shell. Generally one translates these requirements into decay properties in x-space and verifies the latter. Of course life should be simpler in the perturbative setting. We move there now.

Let $\tilde{K}_G(x)$ be the Euclidean x-space kernel of an unrenormalized graph G contributing to a tree τ, as discussed in §2. If we insert the decomposition (B.1) for each propagator and use the formula

$$e^{\alpha_\ell \Delta}(x,y) = (4\pi\alpha_\ell)^{-d/2}\ e^{-(x-y)^2/4\alpha_\ell}$$

for the heat kernel then we find that

$$\tilde{K}_G(x) = \sum_h \int d\alpha\ \chi(\alpha,h) e^{-m^2(\alpha)} \prod_{\ell\in\mathcal{L}(G)} (4\pi\alpha_\ell)^{-d/2}\ e^{-(x_\ell-y_\ell)^2/4\alpha_\ell} . \qquad (B.8)$$

Here $\alpha = (\alpha_\ell)_{\ell \in \mathcal{L}(G)}$, $h = (h_f)_{f \in \mathcal{F}(\tau)}$,

$$\chi(\alpha,h) = \prod_\ell \chi_\ell(\alpha_\ell, h_{f(\ell)})$$

where $f(\ell)$ is the fork at which ℓ first appears (i.e., $\ell \in \mathcal{L}(g_{f(\ell)})$) and

$$\chi_\ell(\alpha_\ell, h_f) = \begin{cases} \chi_{h_f}(\alpha_\ell) & \text{if } \ell \text{ is a hard line} \\ \chi_{<h_f}(\alpha_\ell) & \text{if } \ell \text{ is a soft line} \end{cases}$$

where $\chi_{<h} = \sum_{h'<h} \chi_{h'}$, x_ℓ and y_ℓ are the coordinates of the endpoints of ℓ, and

$$m^2(\alpha) = \sum_\ell \alpha_\ell m_\ell^2 .$$

(We are assuming that there is only one scalar boson present with mass $m > 0$ so that every $m_\ell = m$. For a model like QED involving fermions, there would also be an operator $i\partial\!\!\!/ + m_\ell$ acting on x_ℓ for each fermi line ℓ in (B.8).)

More notation: Let $\beta_\ell = \alpha_\ell^{-1}$ and $\Delta_\ell = x_\ell - y_\ell$ so that the gaussian in (B.8) can be written

$$\prod_\ell e^{-\beta_\ell \Delta_\ell^2/4} = e^{-x\tilde{B}x/4}$$

where the v x v matrix $\tilde{B}$ is singular because of the translation invariance of $\tilde{K}_G$. We break the translation invariance by setting $x_v = 0$ (which we assume is an external vertex) and we introduce

$$K_G(x_1,\ldots,x_{v-1}) = \tilde{K}_G(x_1,\ldots,x_{v-1},0)$$

and the $(v-1)\times(v-1)$ (non-singular) matrix B obtained from $\tilde{B}$ by dropping the vth row and column. Denoting the other v_e-1 external vertices by x^e and the v_i internal vertices by x^i, we now write $x = (x^e,x^i) \in \mathbf{R}^{d(v-1)}$ and $dx = dx_1\ldots dx_{v-1}$. Then the value of the unrenormalized graph can be written

$$G = \int K_G(x)\ \Phi(x^e)\,dx \tag{B.9}$$

where

$$\Phi(x^e) = \int dy \prod_{j=1}^{v_e-1} \phi_j(x_j^e+y)\ \phi_{v_e}(y) .$$

The value of the renormalized graph G_{ren} corresponding to a tree τ with a label R or C at each fork is defined as in §2 by applying Taylor operations at each fork in succession down the tree. At each fork f we take the following steps:

1) introduce an interpolation operation I_f involving a parameter t_f that acts on the external vertices of the subgraph G_f (defined explicitly in (B.11) below);

2) apply a number of t_f-derivatives:

$$T_f = \begin{cases} \partial_{t_f}^{\delta_f+1} & f \in \mathcal{F}_R , \quad \delta_f \geq 0 \\ 1 & f \in \mathcal{F}_R , \quad \delta_f < 0 \\ \partial_{t_f}^{m_f} & f \in \mathcal{F}_C , \quad 0 \leq m_f \leq \delta_f \\ 0 & f \in \mathcal{F}_C , \quad \delta_f < 0 ; \end{cases}$$

3) integrate over t_f with respect to the measure $\rho_f(t_f)dt_f$ where

$$\rho_f(t_f) = \begin{cases} (\delta_f!)^{-1}(1-t_f)^{\delta_f} \chi_{[0,1]}(t_f) & f \in \mathcal{F}_R , \ \delta_f \geq 0 \\ \delta(t_f-1) & f \in \mathcal{F}_R , \ \delta_f < 0 \\ \delta(t_f) & f \in \mathcal{F}_C . \end{cases}$$

These operations are applied inductively down the tree, i.e.,

$$G_{ren} = \int dt\, \rho \prod_f T_f I_f G \tag{B.10}$$

where

$$\int dt\rho = \prod_f \int dt_f \rho_f$$

and where in the product in (B.10), $T_{f'} I_{f'}$ occurs to the right of $T_f I_f$ if $f < f'$.

By a subgraph G_f of G we mean the following: its <u>vertices</u> $\mathcal{V}(G_f)$ are those vertices of G which are associated with endpoints of τ above f; its <u>lines</u> $\mathcal{L}(G_f)$ are the lines of G which join two of its vertices; its <u>legs</u> $\Lambda(G_f)$ are the external fields $\phi_j(x_j)$ depending on a vertex $x_j \in \mathcal{V}(G_f)$ together with the lines of G that join a vertex of G_f to a vertex not in G_f (actually the leg is the "half" of each such line that touches G_f); its <u>external vertices</u> $\mathcal{V}^e(G_f)$ are those of its vertices associated with a leg. We select a vertex $x_f \in \mathcal{V}^e(G_f)$ as the <u>localization vertex</u> for G_f. It is convenient to do so by working up from the bottom of the tree as follows:

i) $x_F = x_V = 0$;

ii) if $x_{\pi(f)} \in \mathcal{V}(G_f)$, take $x_f = x_{\pi(f)}$;

iii) if $x_{\pi(f)} \notin \mathcal{V}(G_f)$, take x_f to be any vertex in $\mathcal{V}^e(G_f) \cap \mathcal{V}^e(G)$ if this set is nonempty, and otherwise take x_f to be any vertex in $\mathcal{V}^e(G_f)$.

(In §2 we denoted the localization vertex by x_f^0, but now we drop the superscript.)

Suppose that $G^{(>f)} \equiv \prod_{f'>f} T_{f'} I_{f'} G$ has been defined. Then $I_f G^{(>f)}$ is defined by replacing the argument x_j of every leg of G_f by

$$x_j(t_{\geq f}) = x_f + t_f(x_j(t_{>f}) - x_f) \tag{B.11}$$

where $t_{>f} = (t_{f'})_{f'>f}$, etc. Note that $x_j(t_{\geq f})$ depends on $t_{>f}$ as well as on t_f since x_j already has a $t_{>f}$-dependence. $x_f(t_{\geq f}) = x_f$ has no t-dependence since we have arranged that x_f is the localization vertex for every $G_{f'}$, $f' > f$, for which it is a vertex. As we move down the tree, each x_j continues to acquire further t-dependence, so long as it remains an external vertex of G_f. At the bottom of the tree, we write $x_j^e(t) = x_j^e(t_{\geq F})$ for the external vertices of G. By our rules i) - iii) above, $x_j^e(t)$ is a t-dependent linear combination of the x_k^e's but is independent of x^i.

When a ∂_{t_f} in T_f acts on $I_f G^{(>f)}$ it acts on one of the legs of G_f; i.e. either on a field by

$$\partial_{t_f} \Phi(x^e(t_{\geq f})) = \partial_{t_f} x_j(t_{\geq f}) \cdot \partial_{x_j} \Phi = (x_j(t_{>f}) - x_f) \cdot \partial_{x_j} \Phi(x^e(t_{\geq f})) \tag{B.12a}$$

or on the coordinate difference of a line ℓ joining $x_j \in \mathcal{V}^e(G_f)$ to $x_k \notin \mathcal{V}(G_f)$ by

$$\partial_{t_f} \Delta_\ell(t_{\geq f}) = x_j(t_{>f}) - x_f . \tag{B.12b}$$

As we continue to move through the product in (B.10) the difference $x_j(t_{>f}) - x_f$ produced in (B.12) acquires no further t-dependence, since it is not the argument of a leg of any $G_{f'}$, $f' < f$. It depends on $t_{>f}$ but not on $t_{\leq f}$, is unaffected by further t-derivatives and is thus an "internal" contribution to $G_{f,\mathrm{ren}}$. On the other hand the fields Φ and coordinate differences $\Delta_\ell(\ell \notin \mathcal{L}(G_f))$

are "external" contributions to $G_{f,\text{ren}}$ and do accumulate further t-dependence. When we finish, $\Phi(x^e(t))$ will depend on all the t_f's and Δ_ℓ will depend on $t_{>f(\ell)}$; but if there is no possible confusion we shall simply write $\Delta_\ell(t)$ instead of $\Delta_\ell(t_{>f(\ell)})$.

Example. Consider the ϕ^4_4 graph G associated with the tree τ:

τ =
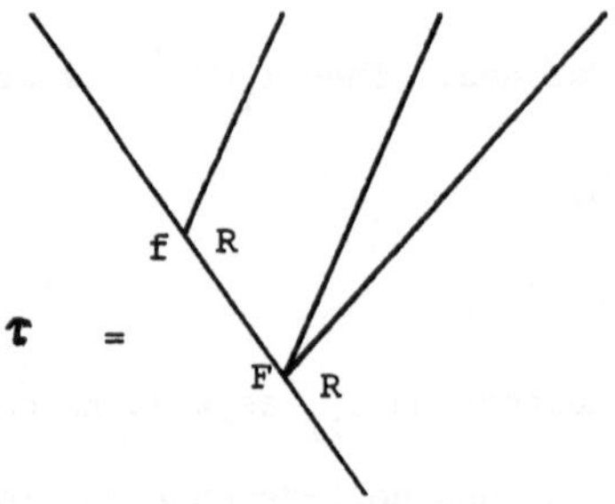

G =
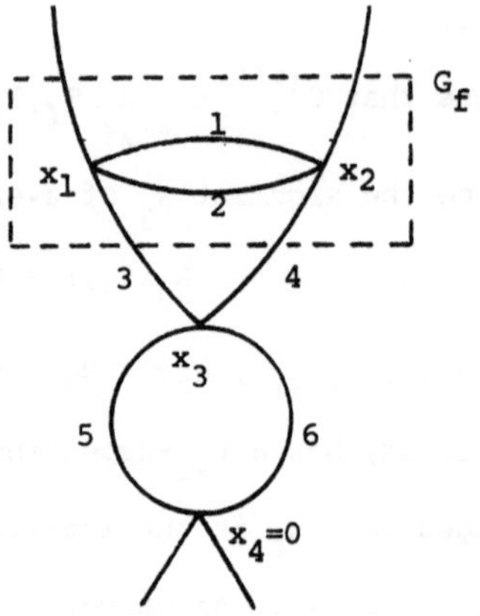

(B.13)

As we follow the definition (B.10) we shall track only the active part of G, i.e. the gaussian and the external fields. Thus we write

$$G \sim \prod_\ell e^{-\beta_\ell \Delta_\ell^2/4} \Phi(x_1,x_2)$$

$$= e^{-\frac{1}{4}\left[\beta_{12}(x_2-x_1)^2 + \beta_3(x_3-x_1)^2 + \beta_4(x_3-x_2)^2 + \beta_{56}x_3^2\right]} \Phi(x_1,x_2)$$

where $\beta_{12} = \beta_1 + \beta_2$ and $\beta_{56} = \beta_5 + \beta_6$. Choosing $x_f = x_1$ we have

$$I_f G \sim \prod_\ell e^{-\beta_\ell \Delta_\ell(t_f)^2/4} \Phi(x_1,x_2(t_f))$$

$$= e^{-\frac{1}{4}\left[\beta_{12}(x_2-x_1)^2 + \beta_3(x_3-x_1)^2 + \beta_4(x_3-x_2(t_f))^2 + \beta_{56}x_3^2\right]} \Phi(x_1,x_2(t_f))$$

where

$$x_2(t_f) = x_1 + t_f(x_2 - x_1) .$$

Then

$$T_f I_f G = \partial_{t_f} I_f G$$

$$\sim \prod_{\ell} e^{-\beta_\ell \Delta_\ell (t_f)^2/4} \Big[\tfrac{1}{2}\, \beta_4 (x_2-x_1)\cdot(x_3-x_2(t_f))\ \Phi(x_1,x_2(t_f))$$

$$+ (x_2-x_1)\ \partial_{x_2}\Phi(x_1,x_2(t_f))\Big] .$$

In terms of

$$x_1(t) = t_F x_1$$

and

$$x_2(t) = t_F x_2(t_f) = t_F x_1 + t_F t_f (x_2-x_1)$$

we have

$$I_F T_f I_f G \sim \prod_{\ell} e^{-\beta_\ell \Delta_\ell (t_f)^2/4} \Big[\tfrac{1}{2}\, \beta_4 (x_2-x_1)\cdot(x_3-x_2(t_f))\ \Phi(x_1(t),x_2(t))$$

$$+ (x_2-x_1)\ \partial_{x_2}\Phi(x_1(t),x_2(t))\Big] .$$

the only legs at this interpolation step being external fields. The final T_F differentiates the fields in the first term (for which $\delta_F = 0$),

$$T_F\Phi(x_1(t),x_2(t)) = x_1\cdot\partial_{x_1}\Phi(x_1(t),x_2(t))$$

$$+ (x_1+t_f(x_2-x_1))\cdot\partial_{x_2}\Phi(x_1(t),x_2(t))$$

but leaves the second term (for which $\delta_F = -1$) unchanged.

Thus each ∂_{t_f} in G_{ren} does one of the following:

a) it acts on an external field as in (B.12a);

b) it brings down a factor $-\beta_\ell \partial_{t_f}\Delta_\ell\cdot\Delta_\ell/2$ from the gaussian; or

c) it acts on a Δ_ℓ already brought down in b).

It follows that

$$\prod_f T_f I_f\ e^{-\Sigma\beta_\ell\Delta_\ell^2/4}\ \Phi = \sum_i R_i\ e^{-\Sigma\beta_\ell\Delta_\ell^2/4}\ \Phi \qquad \text{(B.14)}$$

where each R_i is a monomial in the quantities

$$i\,\partial_{t_f} x_j^e \cdot p_j^e \ , \quad \beta_\ell\, \partial_{t_f} \Delta_\ell \cdot \Delta_\ell \quad \text{and} \quad \beta_\ell\, \partial_{t_f} \Delta_\ell \cdot \partial_{t_{f'}} \Delta_\ell \tag{B.15}$$

where $ip_j^e = \partial_{x_j^e}$ acts on Φ.

We regard each term in the sum (B.14) as leading to a separate renormalized graph, and so the renormalized version of (B.9) is a sum over renormalized graphs of the form

$$G_{ren} = c \sum_h \int dt\rho(t) \int d\alpha\chi(\alpha)\ e^{-m^2(\alpha)} \prod_\ell \beta_\ell^{d/2} \int dx\ Re^{-\mathcal{B}/4}\ \Phi \tag{B.16}$$

where $c = (4\pi)^{-d\ell(G)/2}$, R is a monomial as in (B.15),

$$\mathcal{B}(x,t) = \sum_\ell \beta_\ell \Delta_\ell(t)^2$$

and the sum over scales h respects the convention (2.50). In general we shall use the letter c to denote a general constant which is independent of the variables $x,t,\alpha,h,\ldots$, but which may depend on G, with the understanding that $c \le c_0^{\ell(G)}$, where c_0 is independent of G. If we place the operations ip_j^e in R on Φ to give the external fields

$$\Pi(x^e(t)) = \prod_j \partial_{x_j}^{q_j} \Phi(x^e(t))$$

and write

$$R(x,\alpha,t,p^e) \prod_\ell \beta_\ell^{d/2}\ e^{-\mathcal{B}/4}\ \Phi$$

$$\equiv S(x,\alpha,t) \prod_\ell \beta_\ell^{d/2}\ e^{-\mathcal{B}/4}\ \Pi \equiv K(x,\alpha,t)\ \Pi(x^e(t)) \tag{B.17}$$

then (B.16) takes the form

$$G_{ren} = c \sum \int e^{-m^2(\alpha)} \int dx\ K\ \Pi \tag{B.18}$$

where

$$\sum \int = \sum_h \int dt\rho(t) \int d\alpha\chi(\alpha,h)\ .$$

An important feature of the kernel K is that for each derivative ∂_{t_f} there is a coordinate difference $x_j(t_{>f}) - x_f$ where $x_j \in \mathcal{V}(G_f)$ (see (B.12)).

Let $\mathcal{M}_f = \mathcal{M} \cap \mathcal{L}(g_f)$, $\mathcal{M}_{\geq f} = \bigcup_{f' \geq f} \mathcal{M}_{f'}$, $\mathcal{M}_{>f} = \bigcup_{f'>f} \mathcal{M}_{f'}$. Now we know that x_j is linked to x_f by a chain in $\mathcal{M}_{\geq f}$: for t = 1

$$x_j - x_f = \sum_{\ell \in \mathcal{M}_{\geq f}} u_\ell \Delta_\ell \tag{B.19a}$$

where $u_\ell = 0, \pm 1$. For general t, it is easy to verify by induction that (B.19a) becomes

$$x_j(t_{>f}) - x_f = \sum_{\ell \in \mathcal{M}_{\geq f}} u_\ell(t) \Delta_\ell(t) \tag{B.19b}$$

where

$$u_\ell(t) = u_\ell(t=1)\Theta_{(f,f(\ell)]}$$

and

$$\Theta_{(f_1,f_2]} = \begin{cases} \prod_{f_1<f\leq f_2} t_f & \text{if } f_1<f_2 \\ 1 & \text{if } f_1 \nless f_2 \ . \end{cases} \tag{B.20}$$

As a result we can write the quantities in (B.15) in terms of Δ_ℓ's; for example,

$$\beta_\ell \partial_{t_f} \Delta_\ell \cdot \Delta_\ell = \beta_\ell \sum_{\ell' \in \mathcal{M}_{\geq f}} u_{\ell'}(t) \Delta_{\ell'}(t) \Delta_\ell(t) \ . \tag{B.21}$$

Together with the gaussian $e^{-\mathcal{B}/4}$ in K, each Δ_ℓ leads to a factor $\beta_\ell^{-1/2} = \alpha_\ell^{1/2}$. (B.21), for example, leads to

$$\sum_{\ell' \in \mathcal{M}_{\geq f}} (\alpha_{\ell'}/\alpha_\ell)^{1/2} \ .$$

These are the factors which we used in §2 to show that the kernel has a finite pinned L^1-norm $\|\cdot\|_0$ and which we shall use in Corollary B.4 below.

In order to simplify the complicated t-dependence of the quadratic $\mathcal{B}(x,t)$, we change to a natural set of variables, $x \to \bar{x} = (\bar{x}^e, \bar{x}^i)$, defined as follows:

i) $\bar{x}^e = x^e(t)$;

ii) for $x_j \in \mathcal{V}^i$ we iterate (B.11) all the way down to f = F and set $\bar{x}_j = x_j(t_{\geq F})$.

Note that if f_1 is the highest fork at which $x_j \in \mathcal{V}^i$ is the endpoint of a line, and f_2 is the lowest, then, as an external vertex of G_f, $f_2<f\leq f_1$, x_j depends

naturally on $t_{\geq f}$ and never acquires any dependence on $t_{\leq f_2}$. Nevertheless, we define $\bar{x}_j$ to depend on $t_{\leq f_2}$, in order that $\bar{x}^e$ and $\bar{x}^i$ share a common t-dependence. As a result:

<u>Lemma B.1.</u> a) Let x_j and x_k be two vertices of G_f. Then

$$\bar{x}_j - \bar{x}_k = \Theta_f(x_j(t_{>f}) - x_k(t_{>f})) \tag{B.22a}$$

where

$$\Theta_f = \Theta_{[F,f]} = \prod_{f'\leq f} t_{f'} .$$

In particular, if $\ell = \overrightarrow{x_k x_j} \in \mathcal{L}(g_f)$, then

$$\bar{\Delta}_\ell \equiv \bar{x}_j - \bar{x}_k = \Theta_f \Delta_\ell(t) . \tag{B.22b}$$

b) Letting $\bar{\beta}_\ell = \Theta_{f(\ell)}^{-2}\beta_\ell$, we have

$$\mathcal{B} = \sum_\ell \bar{\beta}_\ell \bar{\Delta}_\ell^2 . \tag{B.23}$$

<u>Proof.</u> a) By (B.11),

$$x_j(t_{>f}) - x_k(t_{>f}) = t_f^{-1}(x_j(t_{\geq f}) - x_k(t_{\geq f})) ,$$

and, iterating down to the bottom of the tree, we obtain (B.22) .

b) This is an immediate consequence of a). ∎

We write the x-integral in (B.18) as

$$\int dx\, K\, \Pi = U(\alpha,t)^{-d/2} \langle S(x,\alpha,t)\, \Pi(x^e(t)) \rangle \tag{B.24a}$$

where

$$\langle \cdot \rangle = \int dx\, e^{-\mathcal{B}/4} \cdot / \int dx\, e^{-\mathcal{B}/4} \tag{B.24b}$$

and

$$U(\alpha,t)^{-d/2} = \Pi\beta_\ell^{d/2} \int dx\, e^{-\mathcal{B}/4} . \tag{B.25}$$

Next we change variables $x \to \bar{x}$ in (B.24). If any $t_f = 0$ the Jacobian will be 0 and $\bar{\beta}_\ell$ will be infinite if $f(\ell) \geq f$. So we assume for now that each $t_f > 0$, but we shall obtain bounds uniform in t as $t_f \to 0$. We obtain

$$\int dx\, K\, \Pi = U^{-d/2} \langle \bar{S}\, \Pi(\bar{x}^e) \rangle \tag{B.26}$$

where the gaussian expectation is given by

$$\langle \cdot \rangle = \int d\bar{x}\, e^{-\mathcal{B}/4} \cdot / \int d\bar{x}\, e^{-\mathcal{B}/4},$$

and, by (B.17) and (B.15), $\bar{S}$ is a monomial in the quantities

$$\partial_{t_f} x_j^e,\ \beta_\ell^{1/2} \partial_{t_f} \Delta_\ell \quad \text{and} \quad \beta_\ell^{1/2} \Delta_\ell \ ,$$

or, by (B.12), in the quantities

$$(x_j(t_{>f}) - x_f),\ \beta_\ell^{1/2}(x_j(t_{>f}) - x_f) \text{ and } \beta_\ell^{1/2} \Delta_\ell \ . \tag{B.27}$$

In the second quantity, $x_j \in \mathcal{V}^e(G_f)$ is the endpoint of the line ℓ on which ∂_{t_f} has acted, whereas in the first quantity $x_j \in \mathcal{V}^e(G_f)$ is the argument of Φ on which ∂_{t_f} has acted. If we index the t-derivatives by ν and let $\ell(\nu)$ denote the leg on which $\partial_{t_{f(\nu)}}$ acts, then we can write the first two terms in (B.27) as

$$\beta_{\ell(\nu)}^{1/2} \ (x_{j(\nu)}(t_{>f(\nu)}) - x_{f(\nu)}) \equiv \beta_{\ell(\nu)}^{1/2} \Delta_\nu$$

where $\beta_{\ell(\nu)} = 1$ if $\ell(\nu)$ is an external field. Going to the barred variables, we see that

$$\bar{S} = W \prod_\nu \Theta_\nu^{-1} \, \bar{\beta}_{\ell(\nu)}^{1/2} \, \bar{\Delta}_\nu \tag{B.28}$$

where W is a monomial in the quantities $\bar{\beta}_\ell^{1/2} \bar{\Delta}_\ell$,

$$\bar{\Delta}_\nu = \bar{x}_{j(\nu)} - \bar{x}_{f(\nu)} \ ,$$

$$\Theta_\nu = \begin{cases} \Theta_{(f(\ell(\nu)),f(\nu)]} & \ell(\nu) \text{ a half-line} \\ \Theta_{f(\nu)} & \ell(\nu) \text{ an external field} \end{cases}$$

and $\bar{\beta}_{\ell(\nu)} = 1$ if $\ell(\nu)$ is an external field.

In (B.26) the factor $\langle \bar{S}\, \Pi \rangle$ gives the contribution of the renormalization derivatives and the factor $U^{-d/2}$ gives the contribution of the unrenormalized graph. From its definition (B.25),

$$U(\alpha,t) = c \prod \alpha_\ell \ |B| \tag{B.29}$$

and satisfies the basic bound:

<u>Lemma B.2.</u> Let $\mathcal{M}^c = \mathcal{L}(G)\setminus\mathcal{M}$. Then

$$U^{-d/2} \le c \prod_{\ell\in\mathcal{M}^c} \beta_\ell^{d/2} . \tag{B.30}$$

<u>Proof.</u> In (B.25) we change from x to "hard-line" variables $\xi = (\xi_\ell)_{\ell\in\mathcal{M}}$ where $\xi_\ell = \Delta_\ell(t)$. The Jacobian of this change of variables is ±1. This follows from the relation

$$\xi_\ell - \Delta_\ell(t=1) = \sum_{\ell'\in\mathcal{M}_{>f(\ell)}} w_{\ell'}(t)\ \xi_{\ell'} \tag{B.31}$$

which can be established by induction. By (B.31), the mapping $\{\Delta_\ell(t=1)\}_{\ell\in\mathcal{M}} \to \xi$ has Jacobian 1. Since $x \to \{\Delta_\ell(t=1)\}$ has Jacobian ±1, the mapping $x \to \xi$ does too. Therefore,

$$U^{-d/2} = \prod \beta_\ell^{d/2} \int d\xi \prod_{\ell\in\mathcal{M}} e^{-\beta_\ell\xi_\ell^2/4} \prod_{\ell\in\mathcal{M}^c} e^{-\beta_\ell\Delta_\ell^2/4}$$

$$\le \prod \beta_\ell^{d/2} \prod_{\ell\in\mathcal{M}} \int d\xi_\ell e^{-\beta_\ell\xi_\ell^2/4}$$

$$= c \prod_{\ell\in\mathcal{M}^c} \beta_\ell^{d/2}. \qquad \blacksquare$$

As for the renormalization factor $\bar{S}$ of (B.28), we have the following bounds:

<u>Lemma B.3.</u> a) The factor $\prod_\nu \bar{\Delta}_\nu$ in (B.28) satisfies

$$|\prod_\nu \bar{\Delta}_\nu|\ e^{-\mathcal{B}/16} \le c \prod_f \bar{\gamma}_f^{\,n_f} \tag{B.32a}$$

where

$$\bar{\gamma}_f = \begin{cases} \Theta_f M^{-h_f} & \text{if } h_f > 0 \\ \bar{\alpha}_f^{\,1/2} & \text{if } h_f = 0 \end{cases} \tag{B.32b}$$

and $\bar{\alpha}_f = \max_{\ell\in\mathcal{M}_f} \bar{\alpha}_\ell$.

b) $$|\langle \bar{S} \rangle| \le c \prod_f M^{-N_f^e(h_f-h_{\pi(f)})}\ \bar{\alpha}_f^{\,n_f/2} \prod_\nu \bar{\alpha}_{\ell(\nu)}^{-1/2} \tag{B.32c}$$

where, as in Theorem 2.5, n_f is the number of renormalization derivatives introduced at f and N_f^e is the total number of renormalization derivatives on the legs of G_f,

$$\tilde{\alpha}_f = \max(1,\alpha_f) \ , \quad \alpha_f = \max_{\ell\in\mathcal{M}_f} \alpha_f \ ,$$

and

$$\tilde{\alpha}_\ell = \max\ (1,\ \alpha_\ell)\ .$$

Remark. In the course of proving (B.32a) we assume that each $t_f > 0$ so that we can divide by $\bar{\gamma}_f$, but at a C-fork we assume that t_f is sufficiently small that

$$\bar{\gamma}_f/\bar{\gamma}_{\pi(f)} \le M^{-1} \tag{B.33}$$

(the left side is proportional to t_f). If f is an R-fork for which $h_f - h_{\pi(f)} \ge 1$, (B.33) is true for all $t_f \le 1$. For a bound like (B.32c) in which there is no t-dependence, we can drop the restriction $t_f > 0$.

Proof. a) Each $\bar{\Delta}_\nu$ can be written as a sum of hard line $\bar{\Delta}_\ell$'s:

$$\bar{\Delta}_\nu = \sum_{\ell\in\mathcal{M}_{\ge f(\nu)}} u_{\nu,\ell}\bar{\Delta}_\ell \ ,$$

where $u_{\nu,\ell} = 0,\ \pm 1$. Letting

$$L_f = \sum_{\ell\in\mathcal{M}_f} |\bar{\Delta}_\ell| \quad \text{and} \quad L_{\ge f} = \sum_{f'\ge f} L_{f'}$$

we then have $|\bar{\Delta}_\nu| \le L_{\ge f(\nu)}$ and

$$\prod_\nu |\bar{\Delta}_\nu| \le \prod_f L_{\ge f}^{n_f} \ .$$

We establish (B.32a) by induction, much as we proved (2.89). To this end we introduce

$$\mathcal{B}_f = \sum_{\ell\in\mathcal{M}_f} \bar{\beta}_\ell \bar{\Delta}_\ell^2$$

and

$$J_f^{(a)} = (a!)^{-1}\,\bar{\gamma}_f^{-a}\,L_{\geq f}^{a} \prod_{f'\geq f} L_{\geq f'}^{n_{f'}}\, e^{-\mathcal{B}_{f'}/16} \tag{B.34}$$

where a = 0, 1, ..., and we make the inductive hypothesis

$$J_f^{(a)} \leq 2^a \prod_{f'\geq f} c_1 c_2^{m_{f'}}\, \bar{\gamma}_{f'}^{-n_{f'}} \tag{B.35}$$

where $m_f = |\mathcal{M}_f|$ and the constants c_1 and c_2 will be specified below. Setting a = 0 and f = F in (B.35) we obtain the bound (B.32a).

Assume now that (B.35) holds at the forks $f_1,\ldots,f_p$ immediately above f (it certainly holds at the endpoints of τ where the induction begins). By the multinomial theorem

$$J_f^{(a)} = (a!)^{-1}\,\bar{\gamma}_f^{-a}\,(L_f + \sum_i L_{\geq f_i})^{a+n_f}\, e^{-\mathcal{B}_f/16} \prod_{f'>f} L_{\geq f'}^{n_{f'}}\, e^{-\mathcal{B}_{f'}/16}$$

$$= \frac{(a+n_f)!}{a!}\,\bar{\gamma}_f^{-a} \sum_{\vec{a}} \frac{L_f^b}{b!}\, e^{-\mathcal{B}_f/16} \prod_i \frac{L_{\geq f_i}^{a_i}}{a_i!} \prod_{f'\geq f_i} L_{\geq f'}^{n_{f'}}\, e^{-\mathcal{B}_{f'}/16} \tag{B.36a}$$

where the sum over $\vec{a} = (a_1,\ldots,a_p)$ takes place over integers $a_i \geq 0$ such that

$$b \equiv a + n_f - \sum a_i \geq 0\ . \tag{B.36b}$$

Now

$$\mathcal{B}_f \geq \sum_{\ell\in\mathcal{M}_f} \bar{\alpha}_f^{-1}\,\bar{\Delta}_\ell^2 \geq 2\lambda\bar{\alpha}_f^{-1/2}\, L_f - \lambda^2 m_f$$

for $\lambda > 0$ an arbitrary constant. Choosing $\lambda = 8M$ if $h_f > 0$ and $\lambda = 8$ if $h_f = 0$ we obtain

$$\mathcal{B}_f/16 \geq \bar{\gamma}_f^{-1}\, L_f - 4M^2\, m_f$$

and so

$$(b!)^{-1}\, L_f^b\, e^{-\mathcal{B}_f/16} \leq c_3^{m_f}\,\bar{\gamma}_f^{-b} = c_3^{m_f}\,\bar{\gamma}_f^{a+n_f-\Sigma a_i}$$

where $c_3 = e^{4M^2}$. Inserting this into (B.36a) and appealing to the definition (B.34), we find that

$$J_f^{(a)} \leq \frac{(a+n_f)!}{a!}\, c_3^{m_f}\,\bar{\gamma}_f^{n_f} \sum_{\vec{a}} \prod_i \bar{\gamma}_f^{-a_i}\,\bar{\gamma}_{f_i}^{a_i}\, J_{f_i}^{(a_i)}\ .$$

By the inductive hypothesis (B.35), and the inequalities (B.33) and

$$\frac{(a+n_f)!}{a!} \le n_f!\; 2^{n_f+a} ,$$

$$J_f^{(a)} \le n_f!\; 2^{n_f+a}\, c_3^{m_f} \prod_{f'>f} c_1 c_2^{m_{f'}} \prod_{f'\ge f} \bar{\gamma}_{f'}^{-n_{f'}} \sum_{\vec{a}} \prod_i (2/M)^{a_i} .$$

If $M \ge 4$ we can drop the restriction (B.36b) on the sum over $\vec{a}$ and bound it by $2^p = 2^{m_f+1}$. Therefore,

$$J_f^{(a)} \le 2^a\, n_f!\; 2^{n_f+1} (2c_3)^{m_f} \prod_{f'>f} c_1 c_2^{m_{f'}} \prod_{f'\ge f} \bar{\gamma}_{f'}^{-n_{f'}} .$$

Choosing $c_2 = 2c_3 = 2e^{4M^2}$ and $c_1 = \max n_f!\; 2^{n_f+1} = 96$ we obtain (B.35) for the fork f.

b) Since any line ℓ occurs at most 3 times in W, we have

$$|W|\, e^{-\mathcal{B}/16} \le c$$

and so by (B.32a)

$$\|\bar{S}e^{-\mathcal{B}/8}\|_\infty \le c \prod_\nu \Theta_\nu^{-1}\, \bar{\beta}_{\ell(\nu)}^{1/2} \prod_f \bar{\gamma}_f^{\,n_f} .$$

$$= c \prod_f M^{-n_f h_f}\, \tilde{\alpha}_f^{\,n_f/2} \prod_\nu \alpha_{\ell(\nu)}^{-1/2} ,$$

the t-dependence having dropped out. Now

$$\alpha_{\ell(\nu)}^{-1/2} \le M^{h_{f(\ell(\nu))}}\, \tilde{\alpha}_{\ell(\nu)}^{-1/2}$$

and so

$$\prod_f M^{-n_f h_f} \prod_\nu \alpha_{\ell(\nu)}^{-1/2} \le \prod_\nu M^{h_{f(\ell(\nu))} - h_{f(\nu)}}\, \tilde{\alpha}_{\ell(\nu)}^{-1/2} .$$

But

$$\sum_\nu (h_{f(\nu)} - h_{f(\ell(\nu))}) = \sum_\nu \sum_{f\in(f(\ell(\nu)),f(\nu)]} (h_f - h_{\pi(f)})$$

$$= \sum_f N_f^e\, (h_f - h_{\pi(f)})$$

and therefore

$$\|\bar{S}e^{-\mathcal{B}/8}\|_\infty \le c \prod_f M^{-N_f^e(h_f - h_{\pi(f)})} \tilde{\alpha}_f^{n_f/2} \prod_\nu \tilde{\alpha}_{\ell(\nu)}^{-1/2} .$$

By scaling the $\bar{x}$'s,

$$|\langle\bar{S}\rangle| \le \frac{\int e^{-\mathcal{B}/8} d\bar{x}}{\int e^{-\mathcal{B}/4} d\bar{x}} \|\bar{S}e^{-\mathcal{B}/8}\|_\infty = 2^{d(v-1)} \|\bar{S}e^{-\mathcal{B}/8}\|_\infty$$

from which we obtain (B.32c). ∎

From (B.18), (B.24), (B.30), and (B.32c) we have

$$|G_{ren}| \le c \, \|\pi\|_\infty \sum \prod_f M^{-N_f^e(h_f - h_{\pi(f)})} \int e^{-m^2(\alpha)} \prod_{\ell\in c} \beta_\ell^{d/2} \prod_f \tilde{\alpha}_f^{n_f/2} \prod_\nu \tilde{\alpha}_{\ell(\nu)}^{-1/2} . \quad (B.37)$$

Discarding the last factor in (B.37), we can now easily integrate out the α_ℓ's provided we use the decay factor $e^{-m^2\alpha_\ell}$ for $\ell \in \mathcal{M}_0$ where

$$\mathcal{M}_0 = \{\ell \in \mathcal{M} | \, h_{f(\ell)} = 0\} .$$

For $\ell \in \mathcal{M}^c$ we integrate out α_ℓ by $(h = h_{f(\ell)})$

$$\int d\alpha_\ell \, \chi_{\le h}(\alpha_\ell) e^{-m^2\alpha_\ell} \beta_\ell^{d/2} = \int_{M^{-2h}}^\infty d\alpha_\ell \, e^{-m^2\alpha_\ell} \alpha_\ell^{-d/2} \le c \, M^{(d-2)h_{f(\ell)}} . \quad (B.38a)$$

For $\ell \in \mathcal{M}\setminus\mathcal{M}_0$,

$$\int d\alpha_\ell \, \chi_h(\alpha_\ell) \le c \, M^{-2h_{f(\ell)}} . \quad (B.38b)$$

For the remaining lines in $\mathcal{M}_0$,

$$\prod_{\ell\in\mathcal{M}_0} \int_1^\infty d\alpha_\ell \, e^{-m^2\alpha_\ell} \prod_f \tilde{\alpha}_f^{n_f/2} \le c . \quad (B.38c)$$

The estimates (B.37) - (B.38) imply that

$$|G_{ren}| \le c \, \|\pi\|_\infty \sum_h \prod_{\ell\in\mathcal{M}^c} M^{(d-2)h_{f(\ell)}} \prod_{\ell\in\mathcal{M}} M^{-2h_{f(\ell)}} \prod_f M^{-N_f^e(h_f - h_{\pi(f)})} .$$

But

$$\sum_{\ell\in\mathcal{M}^c}(d-2)h_{f(\ell)} - \sum_{\ell\in\mathcal{M}} 2h_{f(\ell)} = \sum_{\ell}(d-2)h_{f(\ell)} - d\sum_{\ell\in\mathcal{M}} h_{f(\ell)}$$

$$= \sum_f D^0(g_f)h_f \qquad \text{(see (2.81))}$$

$$= \sum_f D^0(G_f)(h_f-h_{\pi(f)})$$

by summation by parts (Lemma 2.1). Hence we obtain the (easy!) α-x-space version of Theorem 2.5:

Corollary B.4.

$$|G_{ren}| \le c\ \|\pi\|_\infty \sum_h \prod_f M^{(D^0(G_f)-N_f^e)(h_f-h_{\pi(f)})} . \tag{B.39}$$

Remarks

1. The sum over h is performed as in Theorem 2.6.

2. The decay factor $e^{-m^2(\alpha)}$ is not really necessary in (B.38). If $d > 2$, the exponential $e^{-m^2\alpha_\ell}$ is not required in (B.38a). In (B.38c) all that we need is a decay factor α_ℓ^{-a}, $a > 1 + \max_f n_f/2$.

3. In the above estimates, the exponential decay of the Euclidean propagator and of its slices $C^{(h)}$ is a crucial ingredient (see (1.3)). We remark that the (scalar) propagator in relativistic x-space

$$C_R(x) = c\int_0^\infty d\alpha_\ell\ \alpha_\ell^{-d/2}\ e^{-i(\langle x,x\rangle/4\alpha_\ell + \alpha_\ell m^2)}$$

does not have the uniform exponential decay of its Euclidean counterpart. In particular, for $d = 4$, $C_R(x)$ is singular on the light cone $\lambda \equiv \langle x,x\rangle = 0$ and falls off like

$$C_R(x) \sim \begin{cases} |\lambda|^{-3/4}e^{-m|\lambda|^{1/2}} & \lambda \to -\infty \quad \text{(spacelike direction)} \\ |\lambda|^{-3/4} & \lambda \to \infty \quad \text{(timelike direction)} \end{cases} .$$

See, for example, p. 152 of Bogoliubov and Shirkov.[12]

Because of this lack of decay in relativistic x-space and because of the convenient correspondence between a relativistic and a Euclidean graph in p-space, we next transform (B.18) to p-space. We insert

$$\pi(\bar{x}^e) = \int dp^e\, \hat{\pi}(p^e) e^{ip^e \cdot \bar{x}^e}$$

into (B.26), where $dp^e = dp_1^e \ldots dp_{v_e-1}^e$, and apply the basic Fourier transform formula taking $(\bar{x}_1,\ldots,\bar{x}_{v-1})$ to $(p_1,\ldots,p_{v-1})$:

$$\int e^{-\bar{x}\bar{B}\bar{x}/4+i\bar{x}\cdot p}\, d\bar{x} = c\, |\bar{B}|^{-d/2}\, e^{-pAp}$$

where $c = (4\pi)^{d(v-1)/2}$ and $A = \bar{B}^{-1}$. This gives

$$G_{ren} = c \sum \int e^{-m^2(\alpha)}\, U^{-d/2} \int dp^e Q(p^e,\bar{\beta},t)\; e^{-p^e A^e p^e}\; \hat{\pi}(p^e) \tag{B.40}$$

where A^e is the restriction of A to external indices and Q is a polynomial in p^e,

$$Q(p^e,\bar{\beta},t) = \frac{\langle \bar{S} e^{ip^e\cdot\bar{x}^e} \rangle}{\langle e^{ip^e\cdot\bar{x}^e} \rangle} = \sum_j q_j(\bar{\beta},t)\,(p^e)^j \; , \tag{B.41a}$$

whose coefficients $q_j(\bar{\beta},t)$ are rational functions of $\bar{\beta}$ and t. The q_j's are given by

$$q_j = \frac{1}{j!}\, \partial^j_{p^e} \left. \frac{\langle \bar{S} e^{ip^e\cdot\bar{x}^e} \rangle}{\langle e^{ip^e\cdot\bar{x}^e} \rangle} \right|_{p^e=0}$$

$$= \frac{i^{|j|}}{j!}\, \langle \bar{S} | (\bar{x}^e)^j \rangle \tag{B.41b}$$

where $\langle \bar{S} | \bar{x}_{i_1} \ldots \bar{x}_{i_n} \rangle$ denotes the connected expectation in which each $\bar{x}_{i_r}$ contracts with a factor in $\bar{S}$ (given by (B.28)). Note that in (B.41) j is a multi-index,

$$j = (j^\mu_r)\ \begin{matrix} \mu = 1,\ldots,d \\ r = 1,\ldots,v_e-1 \end{matrix}, \quad j^\mu_r \in Z^+ \; ,$$

and that $j! = \prod_{\mu,r} (j^\mu_r)!$.

Remark. The power counting behaviour of (B.41b) is better than might be naively expected. In addition to the factors of $\bar{\gamma}_f$ arising from $\bar{S}$ (see (B.32)), one might anticipate that if the linkage of $\bar{x}^e_k$ to 0 involves a line $\ell' \in \mathcal{M}_0$ then $\bar{x}^e_k$ will contribute a large factor $\alpha^{-1/2}_{\ell'}$ to (B.41b). However, this is not the case since, as we show in (B.58),

$$|\langle\bar{\Delta}_\ell \bar{x}_k^e\rangle| \le 2\,\bar{\alpha}_\ell\ , \qquad (B.42)$$

instead of the expected behaviour $\bar{\alpha}_\ell^{-1/2}\ \bar{\alpha}_{\ell'}^{-1/2}$. Of course, here it is possible that $\ell \in \mathcal{M}_0$ if $f(\ell)$ is a 0-scale C-fork, $\alpha_\ell \ge 1$, and so we are not better off with (B.42). We are though: since $f(\ell)$ is a C-fork, $\bar{\alpha}_\ell = 0$ (a reflection of the decoupling of variables internal and external to a C-fork).

We turn now to the relativistic version $G_{R,ren}$ of G_{ren} of (B.40). We define $G_{R,ren}$ by the same renormalization procedure as in the Euclidean case, except for one minor modification: when the bottom fork F is an R-fork with scale $h_F = 0$ we do not introduce an interpolation parameter t_F in order to attempt a cancellation between 1 and L using Taylor's Theorem, but instead we pull apart R = 1 - L and estimate the 1 and L separately. The reason is that renormalization cancellations are not needed at low scales and in fact are counterproductive: they produce powers of $\tilde{\alpha}_F \ge 1$ (see (B.37)), which are not harmless as in the Euclidean world where there is a decay factor $e^{-m^2(\alpha)}$ (see (B.38c)). From now on we reinterpret the Euclidean expression (B.40) as including this modification as well.

A convenient way to relate G_{ren} and $G_{R,ren}$ is through a "Wick rotation". For a single scalar Euclidean propagator

$$C(p) = \int_0^\infty d\alpha_\ell\ e^{-\alpha_\ell(p^2+m_\ell^2-i\varepsilon)}\ ,$$

where we have added the $i\varepsilon$ ($\varepsilon > 0$) temporarily to improve the large α_ℓ behaviour, a Wick rotation consists of rotating the α_ℓ-contour to the positive imaginary axis:

$$C(p) = i\int_0^\infty d\alpha_\ell\ e^{-i\alpha_\ell(p^2+m_\ell^2-i\varepsilon)}\ .$$

We then make an analytic continuation in $p_0 = p_d$,

$$p(\theta) = (e^{-i\theta}p_0,\vec{p})\ ,\quad \theta : 0 \to \pi/2\ , \qquad (B.43)$$

to obtain the relativistic propagator

$$C(p(\tfrac{\pi}{2})) = i\int_0^\infty d\alpha_\ell \; e^{i\alpha_\ell(\langle p,p\rangle - m_\ell^2 + i\varepsilon)} \equiv C_R(p) \;,$$

where $\langle p,p\rangle = p_0^2 - \vec{p}^2$ is the relativistic norm.

We wish to apply this procedure to the kernel of G_{ren} of (B.40). However, our decomposition function $\chi(\alpha,h)$ (see (B.8)) is non-analytic in α and obstructs the rotation of the α_ℓ-contours. If our decomposition of G had been based on a set of functions $\chi_H(\alpha,h)$ which involved only order relations among the α_ℓ's (as in the decomposition into Hepp sectors[4]) there would have been no obstruction to such a rotation: then we would change variables $(\alpha_1,\ldots,\alpha_L) \to (\lambda,\tilde{\alpha}_1,\ldots,\tilde{\alpha}_{L-1})$ where $\lambda = \alpha_1 + \ldots + \alpha_L$ and $\tilde{\alpha}_j = \lambda^{-1}\alpha_j$. Since U, q_j and A^e are all homogeneous in α (A^e having degree 1) and since

$$\chi_H(\alpha,h) = \chi_H(\tilde{\alpha},h) \;, \tag{B.44}$$

the integral over a single $\chi_H(\alpha,h)$ of a monomial $q_j(p^e)^j$ in Q becomes

$$\int d\tilde{\alpha}\; \tilde{\chi}_H\; \tilde{U}^{-d/2}\; \tilde{q}_j \int_0^\infty d\lambda\; \lambda^a \int dp^e e^{-\lambda(p^e\tilde{A}^e p^e + \tilde{m}^2 - i\varepsilon)} (p^e)^j\; \hat{\pi}(p^e) \tag{B.45}$$

where $a > -1$, and $\tilde{\chi}_H = \chi_H(\tilde{\alpha},h)$, $\tilde{U} = U(\tilde{\alpha},t)$, etc. The λ-integral can now be rotated to the positive imaginary axis to become

$$i^{a+1+|k|} \int_0^\infty d\lambda\; \lambda^a \int dp^e e^{-i\lambda(p^e\tilde{A}^e p^e + \tilde{m}^2 - i\varepsilon)} (p^e)^{j+k}\; \hat{\Phi}(p^e)$$

where we have substituted $\hat{\pi}(p^e) = (ip^e)^k\, \hat{\Phi}^e(p^e)$. The kernel of $\hat{\Phi}^e(p^e)$ in (B.45) thus takes the form

$$i^b \int d\tilde{\alpha}\; \tilde{\chi}_H\; \tilde{U}^{-d/2}\; \tilde{q}_j \int_0^\infty d\lambda\; \lambda^a\; e^{-i\lambda(p^e\tilde{A}^e p^e + \tilde{m}^2 - i\varepsilon)} (p^e)^{j+k} \tag{B.46}$$

where $b = a+1+|k|$.

Now, defining $p_j^e(\theta)$ as in (B.43) we see that (B.46) is an analytic function of θ in a complex neighbourhood of $(0,\pi/2)$, since

$$\mathrm{Re}\; ip^e(\theta)\; \tilde{A}^e p^e(\theta) \geq 0 \quad \text{for } 0 \leq \theta \leq \pi/2 \;.$$

Hence we can analytically continue (B.46) from $\theta = 0$ to $\theta = \pi/2$ to obtain the relativistic kernel

$$i^b \int d\tilde{\alpha}\, \tilde{\chi}_H\, \tilde{U}^{-d/2}\, \tilde{q}_j \int_0^\infty d\lambda\, \lambda^a\, e^{i\lambda(\langle p^e,\tilde{A}^e p^e\rangle - \tilde{m}^2 + i\varepsilon)}\, p^e \left(\frac{\pi}{2}\right)^{j+k} \qquad (B.47a)$$

or, expressed back in terms of the original α-parameters,

$$i^b \int d\alpha\, \chi_H\, U^{-d/2}\, q_j\, e^{i(\langle p^e,A^e p^e\rangle - m^2 + i\varepsilon\lambda)}\, p^e \left(\frac{\pi}{2}\right)^{j+k} . \qquad (B.47b)$$

Of course, in (B.47a) the λ-integral can be evaluated to give

$$(B.47a) = i^{a+b+1}\, a! \int d\tilde{\alpha}\, \tilde{\chi}_H\, \tilde{U}^{-d/2}\, \tilde{q}_j\, [\langle p^e,\tilde{A}^e p^e\rangle - \tilde{m}^2 + i\varepsilon]^{-a-1}\, p^e \left(\frac{\pi}{2}\right)^{j+k}. \qquad (B.47c)$$

On the basis of this formula, the return to real time amounts to controlling

$$\lim_{\varepsilon\to 0+} [\langle p^e,\tilde{A}^e p^e\rangle - \tilde{m}^2 + i\varepsilon]^{-a-1}$$

as a distribution in p^e (uniformly in $\tilde{\alpha}$). This is the approach taken by Hepp in his original paper[4] and by Lowenstein and Speer[36] in their extension of Hepp's results to the case of zero mass particles.

Unfortunately, since our χ's are not scale invariant as in (B.44), this approach is not available to us, nor can we, strictly speaking, Wick rotate and analytically continue the α-p kernels from the Euclidean to the relativistic world. Thus we define $G_{R,ren}$ from first principles as described above. Nevertheless, we find it convenient to imagine that $G_{R,ren}$ has been obtained from G_{ren} by a formal Wick rotation and analytic continuation because it is so easy to write down $G_{R,ren}$ in this way, given G_{ren}. A formal rotation of (B.40) produces

$$G_{R,ren} = c \sum \int e^{-im^2(\alpha)-\varepsilon\lambda}\, U^{-d/2} \sum_j q_j \int dp^e\, e^{i\langle p^e,A^e p^e\rangle}\, (p^e)^{j+k}\, \hat{\Phi}(p^e) \qquad (B.48)$$

where the factors of i have been absorbed into the constant c and where $\varepsilon > 0$ is included to guarantee convergence of the α_ℓ-integrals at ∞ but will be set equal to 0 in the proof of Theorem B.9 below.

The key point is that the functions $U(\bar{\beta},t)$, $q_j(\bar{\beta})$ and $A^e(\bar{\beta})$ are common to (B.40) and (B.48). As a result we can control $G_{R,ren}$ largely by Euclidean x-space estimates. The main distinction between the two formulas is that the Euclidean decay factors $e^{-m^2(\alpha)}$ and $e^{-p^e A^e p^e}$ become oscillatory in the relativistic formula. As we noted in Remark 2 after Corollary B.4 it suffices to have the decay factor α_ℓ^{-a}, $a > 1 + \max_f n_f/2$, for each $\ell \in \mathcal{M}_0$. Now a line of $\mathcal{M}_0$ may occur at a C-fork or, and this is the only possibility for an R-fork, at the bottom fork F when the root scale is -1 (and F is an R-fork with $h_F = 0$). Recall that in the latter case we pull apart the operation $R = 1 - L$ and treat the 1- and L-forks separately. In the case of a C- or the L-fork (we denote these $\mathcal{M}_0$-lines as $\mathcal{M}_{0,C}$), $n_f \le \delta_f \le d - 2\,\frac{d-2}{2} = 2$; therefore we need an additional decay factor

$$\alpha_\ell^{-a}\ , \quad a > 2, \quad \text{for } \ell \in \mathcal{M}_{0,C}\ . \tag{B.49a}$$

In the case of the 1-fork (we denote these $\mathcal{M}_0$ lines by $\mathcal{M}_1$), $n_F = 0$; therefore we need the additional factor

$$\alpha_\ell^{-a}\ , \quad a > 1, \quad \text{for } \ell \in \mathcal{M}_1\ . \tag{B.49b}$$

We extract the decay factors (B.49) by integrating by parts with respect to α_ℓ in (B.48), each integration by parts producing a factor of $\alpha_\ell^{-1} = \beta_\ell$. To facilitate this procedure we modify the partition functions χ_0 and χ_1 (see (B.2)) by smoothing them out at their interface. That is, we choose $\tilde{\chi}_0 \in C^\infty$ with $\operatorname{supp} \tilde{\chi}_0 \subset [1,\infty)$ and $\tilde{\chi}_0(\alpha) \equiv 1$ for $\alpha_\ell \ge 2$, and we choose $\tilde{\chi}_1$ so that

$$\tilde{\chi}_0 + \tilde{\chi}_1 = \chi_0 + \chi_1 = \text{characteristic function of } [M^{-2},\infty)\,.$$

For $\ell \in \mathcal{M}_{0,C}$ we perform the α_ℓ-integration by parts in (B.48) in x-space before taking t-derivatives and Fourier transforming, whereas in the the case $\ell \in \mathcal{M}_1$ we integrate by parts afterwards in p-space. Of course, the results are the same whether the integration by parts occurs before or after the t-derivatives and Fourier transform, but the representations obtained are different. At a C- or L-fork the decoupling between internal and external variables permits us to introduce coordinate differences Δ_ℓ with impunity, whereas at the 1-fork we must exercise more care.

Consider first the case of a line $\ell \in \mathcal{M}_{0,C}$ where the integration by parts is carried out in relativistic x-space. The modified 0-scale propagator is

$$\tilde{C}^{(0)}(\Delta_\ell(t)) = c \int_0^\infty d\alpha_\ell\, \tilde{\chi}_0(\alpha_\ell)\, \beta_\ell^{d/2} e^{iF_\ell}$$

where

$$F_\ell = -\beta_\ell \langle\Delta_\ell(t),\Delta_\ell(t)\rangle/4 - \alpha_\ell(m^2-i\varepsilon) .$$

Integrating by parts gives

$$\tilde{C}^{(0)}(\Delta_\ell) = \frac{c}{i(m^2-i\varepsilon)} \int_0^\infty d\alpha_\ell\, e^{-i\alpha_\ell(m^2-i\varepsilon)} \partial_{\alpha_\ell}(\tilde{X}_0\beta_\ell^{d/2} e^{-i\beta_\ell\langle\Delta_\ell,\Delta_\ell\rangle/4}) ,$$

and so, since $\partial_{\alpha_\ell}\beta_\ell = -\beta_\ell^2$,

$$\tilde{C}^{(0)}(\Delta_\ell) = \frac{c}{\varepsilon+im^2} \int_0^\infty d\alpha_\ell\, (\tilde{X}_0' - \frac{d}{2}\tilde{X}_0\,\beta_\ell + \frac{i}{4}\tilde{X}_0\beta_\ell^2 \langle\Delta_\ell,\Delta_\ell\rangle)\, \beta_\ell^{d/2}\, e^{iF_\ell} .$$

The first term, with support in [1,2], certainly supplies a factor β_ℓ^3 at $\alpha_\ell = \infty$; the second term has an additional factor of β_ℓ and the third term in effect has an additional β_ℓ. So we integrate by parts again (twice) in the last two terms, until we obtain

$$\tilde{C}^{(0)} = \sum_{j=0}^{3} \tilde{C}_j^{(0)} \equiv \int_0^\infty d\alpha_\ell \sum_{j=0}^{3} \zeta_j(\alpha_\ell)(\beta_\ell\langle\Delta_\ell,\Delta_\ell\rangle)^j\, \beta_\ell^{d/2+3} e^{iF_\ell} \tag{B.50}$$

where the cutoff functions $\zeta_j \in C^\infty$ with supp $\zeta_j \subset [1,\infty)$ and ζ_j bounded.

We then insert the sum (B.50) into $G_{R,ren}$ for each line $\ell \in \mathcal{M}_{0,C}$, and, as usual, we make a choice of one of the four possibilities at each ℓ. Suppose $\tilde{C}_1^{(0)}$ is chosen. Then $G_{R,ren}$ contains an additional factor $c\beta_\ell^4 \langle\Delta_\ell(t),\Delta_\ell(t)\rangle$. Of course this factor may be acted upon by t_f-derivatives from forks f higher up the tree and so $G_{R,ren}$ contains one of

$$\beta_\ell^4\langle\Delta_\ell,\Delta_\ell\rangle , \quad \beta_\ell^4\langle\partial_{t_f}\Delta_\ell,\Delta_\ell\rangle \quad\text{or}\quad \beta_\ell^4\langle\partial_{t_f}\Delta_\ell,\partial_{t_{f'}}\Delta_\ell\rangle .$$

Comparing with the (Euclidean) terms in (B.15), we see that except for an extra β_ℓ^3 the last two terms may already occur in $G_{R,ren}$. As a result the choice of $\tilde{C}_1^{(0)}$ produces as an extra factor, one of $\beta_\ell^4\langle\Delta_\ell,\Delta_\ell\rangle$ or β_ℓ^3. The choice of any of the other $\tilde{C}_j^{(0)}$'s produces a similar result, namely, an inconsequential change in the cutoff function and an additional factor $\beta_\ell^3(\beta_\ell\langle\Delta_\ell(t),\Delta_\ell(t)\rangle)^r$ where $r = 0,1,2$ or 3.

The effect of the substitution (B.50) on the value (B.48) of $G_{R,ren}$ is to modify the coefficients

$$q_j = \frac{i^{|j|}}{j!} \langle \bar{S} | (\bar{x}^e)^j \rangle \qquad \text{(B.51a)}$$

(recall that this expectation is Euclidean) by introducing a factor

$$\beta_\ell^3 (\bar{\beta}_\ell \bar{\Delta}_\ell^{-2})^r, \quad r = 0,1,2 \text{ or } 3, \qquad \text{(B.51b)}$$

into $\bar{S}$. Dominating $\bar{\Delta}_\ell$ by the gaussian as in (B.35), we can heuristically regard $\bar{\beta}_\ell \bar{\Delta}_\ell^{-2}$ as ≈ 1 and so we expect to obtain the additional factor β_ℓ^3.

Next consider a line $\ell \in \mathcal{M}_1$ occurring at the bottom 1-fork. Integration by parts in the α_ℓ-integral in (B.48) yields ($\partial = \partial/\partial\alpha_\ell$)

$$\int_0^\infty d\alpha_\ell \, \tilde{X}_0(\alpha_\ell) \, e^{-i\alpha_\ell(m^2 - i\varepsilon)} \sum_j q_j \, U^{-d/2} \int dp^e \, e^{i\langle p^e, A^e p^e \rangle} (p^e)^{j+k} \hat{\Phi}(p^e)$$

$$= \frac{1}{i(m^2 - i\varepsilon)} \int_0^\infty d\alpha_\ell \, e^{-i\alpha_\ell(m^2 - i\varepsilon)} \sum_j U^{-d/2} \int dp^e \, e^{i\langle p^e, A^e p^e \rangle} (p^e)^{j+k} \hat{\Phi}(p^e)$$

$$[\tilde{X}_0' \, q_j + \tilde{X}_0 \, \partial q_j - \tfrac{d}{2} \tilde{X}_0 \, q_j \frac{\partial U}{U} + \tilde{X}_0 \, q_j \, i\langle p^e, \partial A^e p^e \rangle] \, . \qquad \text{(B.52)}$$

As it stands, the last term in (B.52) does not exhibit additional decay in α_ℓ and so we integrate by parts with respect to p^e (twice) via the formula

$$\int dp^e \, i\langle p^e, \partial A^e p^e \rangle \, e^{i\langle p^e, A^e p^e \rangle} F(p^e)$$

$$= \tfrac{1}{2} \int dp^e \left[\langle p^e, (\partial A^e)(A^e)^{-1} \partial_p e \rangle \, e^{i\langle p^e, A^e p^e \rangle} \right] F(p^e)$$

$$= -\tfrac{d}{2} \, \mathrm{tr}((\partial A^e)(A^e)^{-1}) \int dp^e \, e^{i\langle p^e, A^e p^e \rangle} F(p^e)$$

$$+ \tfrac{i}{4} \int dp^e \, e^{i\langle p^e, A^e p^e \rangle} \langle \partial_p e, \, \partial(A^e)^{-1} \partial_p e \rangle \, F(p^e) \, .$$

Since

$$\mathrm{tr}((\partial A^e)(A^e)^{-1}) = \partial |A^e| / |A^e|$$

the integral over p^e in (B.52) can be written as

$$\int dp^e\, e^{i\langle p^e, A^e p^e\rangle} \Big[\tilde{X}_0' q_j + \tilde{X}_0\, \partial q_j - \frac{d}{2} \tilde{X}_0 q_j\, \partial \log(U\cdot|A^e|) + \frac{i}{4} \tilde{X}_0 q_j \langle \partial_p e, \partial(A^e)^{-1} \partial_p e\rangle \Big] (p^e)^{j+k} \hat{\Phi}(p^e) \ . \qquad (B.53)$$

A second integration by parts with respect to α_ℓ will be required to produce the desired factor β_ℓ^2. First, however, we establish the enhanced decay of the differentiated quantities ∂q_j, $\partial \log(U\cdot|A^e|)$ and $\partial(A^e)^{-1}$ in (B.53). The basic tool we use is a formula for the covariance $\langle x_i x_j\rangle$ that we state in the case $t = 1$ (but which clearly also holds when all quantities are barred):

<u>Lemma B.5.</u> Let $\langle\cdot\rangle$ denote the normalized expectation with gaussian density $e^{-\mathcal{B}/4} = e^{-xBx/4}$ in the case $t = 1$.

a) $$\langle x_i x_j\rangle = 2B_{ij}^{-1} = 2U^{-1} \sum_{S\in \mathcal{S}_{ij}(G)} \beta_S \qquad (B.54a)$$

where

$$U = |B| = \sum_{T\in \mathcal{T}(G)} \beta_T \ . \qquad (B.54b)$$

Here $\mathcal{T}(G)$ is the set of trees connecting the vertices of G, $\mathcal{S}_{ij}(G)$ is the set of trees minus one line ("2-trees") for which the vertices x_i and x_j are disconnected from the vertex x_v, and

$$\beta_T = \prod_{\ell\in T} \beta_\ell \ . \qquad (B.55)$$

b) If $\ell = \overrightarrow{x_i x_j}$ then

$$\langle \Delta_\ell x_k\rangle = 2\alpha_\ell\, U^{-1} \sum_T u_{k,\ell,T}\, \beta_T \qquad (B.56)$$

where

$$u_{k,\ell,T} = \begin{cases} 1 & \text{if } T\backslash\ell \text{ joins } x_i \text{, but not } x_j \text{ or } x_k \text{, to } x_v \\ -1 & \text{if } T\backslash\ell \text{ joins } x_j \text{, but not } x_i \text{ or } x_k \text{, to } x_v \\ 0 & \text{otherwise .} \end{cases}$$

<u>Proof.</u> a) This is a standard result and may be found, for example, in (8-17) of Nakanishi's text[37].

b) (B.56) follows from (B.54), once we check that

$$\sum_{S\in \mathcal{S}_{jk}} \beta_S - \sum_{S\in \mathcal{S}_{ik}} \beta_S = \sum_T u_{k,\ell,T}\, \beta_{T\backslash\ell} \ . \qquad (B.57)$$

On the left side of (B.57) the 2-trees containing ℓ cancel since if $\ell \in S$ then

$S \in \mathcal{S}_{jk} \longleftrightarrow S \in \mathcal{S}_{ik}$. If $\ell \notin S$ then

$$S \in \mathcal{S}_{jk} \ominus \mathcal{S}_{ik} \rightarrow S + \ell \in \mathcal{T} .$$

For if $S + \ell \notin \mathcal{T}$ then S must join x_i and x_j and hence cannot be in $\mathcal{S}_{jk} \ominus \mathcal{S}_{ik}$. (B.57) follows from these two facts. ■

According to our method of bounding $\beta_\ell^{1/2}\Delta_\ell$ by the gaussian,

$$|\langle \Delta_\ell \Delta_{\ell'} \rangle| \leq c\, \alpha_\ell^{1/2}\, \alpha_{\ell'}^{1/2}$$

and

$$|\langle \Delta_\ell x_k \rangle| \leq c\, \alpha_\ell^{1/2} \max_{\ell' \in \mathcal{C}} \alpha_{\ell'}^{1/2}$$

where $\mathcal{C}$ is a chain of lines joining x_k to $x_v = 0$. These estimates are, however, too naive when $\alpha_{\ell'} > 1$:

<u>Corollary B.6.</u>

$$|\langle \Delta_\ell x_k \rangle| \leq 2\, \alpha_\ell \tag{B.58a}$$

and

$$|\langle \Delta_\ell \Delta_{\ell'} \rangle| \leq 4 \min (\alpha_\ell , \alpha_{\ell'}) . \tag{B.58b}$$

<u>Proof.</u> (B.58a) is an immediate consequence of (B.56) and (B.54b). (B.58b) follows from (B.58a): if $\alpha_{\ell'} \geq \alpha_\ell$ we write $\Delta_{\ell'} = x_{j'} - x_{k'}$ and apply (B.58a). ■

At this point we abandon the goal of obtaining the best possible constants in our estimates. The subsequent bounds will involve constants $\kappa = \kappa(G)$ whose G-dependence may be worse than $c = c_0^{\ell(G)}$. Presumably, it is possible to replace κ with c if one takes more care with the combinatorics and does not discard powers of t_f as we do.

We first bound $\partial_{\mathcal{M}_2} q_j \equiv \prod_{\ell \in \mathcal{M}_2} \partial_{\alpha_\ell} q_j$ where q_j is the coefficient (B.51) in $G_{R,ren}$ and $\mathcal{M}_2$ is a subset of $\mathcal{M}_1^2$, where a line may occur in $\mathcal{M}_2$ with multiplicity 1 or 2:

Lemma B.7.

$$|\partial_{\mathcal{M}_2} q_j| \le \kappa\, \beta^3_{\mathcal{M}_{0,C}}\, \beta_{\mathcal{M}_2} \prod_f \tilde{\alpha}_f^{\,n_f/2}\, M^{-N^e_f(h_f - h_{\pi(f)})} \tag{B.59}$$

where $\beta_S = \prod_{\ell \in S} \beta_\ell$ (including multiplicities in the case $S = \mathcal{M}_2$).

Proof. By (B.51) and (B.28)

$$q_j = c\, \beta^3_{\mathcal{M}_{0,C}} \langle W \prod_\nu \theta_\nu^{-1}\, \bar{\beta}^{-1/2}_{\ell(\nu)} \bar{\Delta}_\nu | (\bar{x}^e)^j \rangle \tag{B.60}$$

where the monomial W includes the factors $\bar{\beta}_\ell \bar{\Delta}_\ell^{-2}$ generated by the derivatives ∂_{α_ℓ}, $\ell \in \mathcal{M}_{0,C}$ (see (B.51b)), in addition to the renormalization factors $\beta^{1/2}_{\ell(\nu)}\, \Delta_{\ell(\nu)}$ (see B.27). As in the proof of Lemma B.3 we write each $\bar{\Delta}_\nu$ as a sum of $\bar{\Delta}_\ell$'s where $\ell \in \mathcal{M}_{\ge f(\nu)}$, and we consider a typical term in (B.60) where each $\bar{x}^e_k$ contracts with a $\bar{\Delta}_{\ell_k}$ on the left. By (B.58a)

$$|\langle \bar{\Delta}_{\ell_k} \bar{x}^e_k \rangle| \le 2\bar{\alpha}_{\ell_k} . \tag{B.61}$$

We wish to replace $\bar{\alpha}_{\ell_k}$ by $\bar{\alpha}^{-1/2}_{\ell_k}$ so that each $\bar{\Delta}_\ell$ in (B.60) contributes a factor $\bar{\alpha}_\ell^{-1/2}$. If $\bar{\alpha}_{\ell_k} \le 1$ then of course $\bar{\alpha}_{\ell_k} \le \bar{\alpha}^{-1/2}_{\ell_k}$. The possibility $\bar{\alpha}_{\ell_k} > 1$ occurs

a) if $\ell_k = \ell(\nu_k)$ is a soft line contributing to W; or

b) if ℓ_k is a 0-scale hard line (necessarily within a C-fork).

In case a) we have

$$\bar{\alpha}_{\ell_k} \le \tilde{\alpha}^{\,1/2}_{\ell(\nu_k)}\, \bar{\alpha}^{-1/2}_{\ell_k} \tag{B.62a}$$

since $\tilde{\alpha}_\ell = \max(1, \alpha_\ell) \ge \bar{\alpha}_\ell$. By the method of Lemma B.3 we then bound the term under consideration by

$$c \prod_{\ell \in \mathcal{L}_b} \bar{\alpha}_\ell^{-1/2}\, \beta^3_{\mathcal{M}_{0,C}} \prod_f \tilde{\alpha}_f^{\,n_f/2}\, M^{-N^e_f(h_f - h_{\pi(f)})} \tag{B.62b}$$

where $\mathcal{L}_b$ is the set of lines arising in case b) and where the factors $\tilde{\alpha}^{\,1/2}_{\ell(\nu_k)}$ in (B.62a) are cancelled by the last factor in (B.32c). Now set $t_f = C$ at every C-fork f. The only t-dependence in (B.62b) occurs in the first product, which vanishes if $\mathcal{L}_b \ne \emptyset$. Thus we have established (B.59) when $\mathcal{M}_2 = \emptyset$.

Next consider the effect of a derivative ∂_{α_ℓ} ($\ell \in \mathcal{M}_2$) on q_j. If ℓ occurs in $\bar{S}$ so that there is a factor of $\bar{\beta}_\ell = \beta_\ell$ in the product on the left in (B.60), then ∂_{α_ℓ} can act on β_ℓ to give $-\beta_\ell^2$, i.e. an additional factor of β_ℓ. Alternatively it can act on one of the factors $\langle \bar{\Delta}_{\ell_k} \bar{x}_k^e \rangle$ or $\langle \bar{\Delta}_{\ell'}, \bar{\Delta}_{\ell''} \rangle$ contributing to q_j. Now ($\Delta_\ell = \bar{\Delta}_\ell$)

$$|\partial_{\alpha_\ell} \langle \bar{\Delta}_{\ell_k} \bar{x}_k^e \rangle| = \beta_\ell^2 \; |\langle \bar{\Delta}_{\ell_k} \bar{x}_k^e;\ \Delta_\ell^2 \rangle|/4$$

$$= \beta_\ell^2 \; |\langle \bar{\Delta}_{\ell_k} \bar{\Delta}_\ell \rangle \; \langle \bar{x}_k^e \; \bar{\Delta}_\ell \rangle|/2$$

$$\leq 4\beta_\ell^2 \; \bar{\alpha}_{\ell_k} \; \alpha_\ell \qquad \text{(by (B.58))}$$

$$= 2\beta_\ell \; \bar{\alpha}_{\ell_k} \; .$$

Referring back to (B.61) we see that the derivative has produced an extra factor of β_ℓ. The action of ∂_{α_ℓ} on $\langle \bar{\Delta}_{\ell'}, \bar{\Delta}_{\ell''} \rangle$ or of higher derivatives is similar, each ∂_{α_ℓ} producing a factor of β_ℓ (in many ways). In this way we obtain (B.59) where the constant $\kappa = \kappa(G)$ may be quite large as a result of the many terms produced by the contractions and derivatives. ∎

Next we consider the term $\partial_{\alpha_\ell} (A^e)^{-1}$ in (B.53):

<u>Lemma B.8.</u> For $\mathcal{M}_2 \subset \mathcal{M}_1^2$

$$|\partial_{\mathcal{M}_2} (A^e)^{-1}_{ij}| \leq \kappa \; \beta_{\mathcal{M}_2} \; . \qquad \text{(B.63)}$$

<u>Proof.</u> We have

$$(A^e)^{-1}_{ij} = 2 \int p_i^e p_j^e \; e^{-p^e A^e p^e} \; dp^e / \int e^{-p^e A^e p^e} \; dp^e$$

$$= -2 \int \frac{\partial^2}{\partial \bar{x}_i^e \; \partial \bar{x}_j^e} \; e^{-\mathcal{B}/4} \Big|_{\bar{x}^e = 0} \; d\bar{x}^i / \int e^{-\mathcal{B}^i/4} \; d\bar{x}^i$$

where $\mathcal{B}^i = \mathcal{B}|_{\bar{x}^e = 0}$. Now

$$\frac{\partial}{\partial \bar{x}_i^e} \mathcal{B} = 2(E\bar{\beta}\bar{\Delta})_i = 2 \sum_\ell E_{i\ell}\bar{\beta}_\ell\bar{\Delta}_\ell$$

where E is the incidence matrix

$$E_{i\ell} = \begin{cases} 1 & \text{if } x_i^e \text{ is the final point of } \ell \\ -1 & \text{if } x_i^e \text{ is the initial point of } \ell \\ 0 & \text{otherwise .} \end{cases} \tag{B.64}$$

Thus we obtain the following formula for $(A^e)^{-1}$:

$$(A^e)^{-1}_{ij} = (E\bar{\beta}E^t)_{ij} - \frac{1}{2} \langle (E\bar{\beta}\bar{\Delta})_i (E\bar{\beta}\bar{\Delta})_j \rangle^i \tag{B.65}$$

where

$$\langle F(\bar{x}) \rangle^i = \int F(\bar{x}) \Big|_{\bar{x}^e=0} e^{-\mathcal{B}^i/4} \, d\bar{x}^i / \int e^{-\mathcal{B}^i/4} \, d\bar{x}^i \, .$$

Now apply ∂_{α_ℓ} to (B.65). If β_ℓ occurs explicitly in (B.65) the derivative ∂_{α_ℓ} can act on it to give a β_ℓ^2; e.g., if $E_{i\ell} = \pm 1$ then

$$|\langle (\partial_{\alpha_\ell} E_{i\ell}\beta_\ell)\bar{\Delta}_\ell \ (E\bar{\beta}\bar{\Delta})_j \rangle^i| = \beta_\ell^2 \ |\sum_{\ell'} E_{j\ell'} \ \bar{\beta}_{\ell'} \ \langle \bar{\Delta}_\ell \bar{\Delta}_{\ell'} \rangle^i|$$

$$\le c \ \beta_\ell^2$$

by (B.58b) since $\Delta_\ell = \bar{\Delta}_\ell$. Alternatively, ∂_{α_ℓ} can act on $\mathcal{B}^i$ to produce

$$\beta_\ell^2 |\langle (E\bar{\beta}\bar{\Delta})_i (E\bar{\beta}\bar{\Delta})_j ; \ \Delta_\ell^2 \rangle| / 8$$

$$= \beta_\ell^2 |\langle (E\bar{\beta}\bar{\Delta})_i \bar{\Delta}_\ell \rangle \langle (E\bar{\beta}\bar{\Delta})_j \bar{\Delta}_\ell \rangle| / 4 \le c \ \beta_\ell^2$$

by (B.58b). Each subsequent $\alpha_{\ell'}$-derivative produces an additional factor $\beta_{\ell'}$.

∎

The other quantities differentiated in (B.53) are

$$\log U^{-d/2} = \frac{d}{2} \sum_\ell \log \beta_\ell + \log \int dx \ e^{-\mathcal{B}/4} \, , \tag{B.66a}$$

(see (B.25)), and

$$\log |A^e|^{-d/2} = c + \log \int e^{-\mathcal{B}^i/4} \, d\bar{x}^i \, . \tag{B.66b}$$

It follows easily from (B.66) that

$$|\partial_{\mathcal{M}_5} \log (U|A^e|)| \le c\,\beta_{\mathcal{M}_5} . \tag{B.67}$$

Finally we arrive at the bound on a graph of the relativistic theory. Recall that we let $\hat{\Pi}(p^e) = (p^e)^k\,\hat{\Phi}(p^e)$ where $(p^e)^k$ results from the renormalization derivatives acting on the external fields Φ .

<u>Theorem B.9.</u> Let $G_{R,ren}$ of (B.48) be a renormalized graph arising in the tree expansion for the effective potential $V_r(\Phi)$ of a relativistic field theory of a massive scalar field with dimensionless interaction (r is the root scale).

a) If the bottom fork F is an R-fork and $r \ge 0$,

$$|G_{R,ren}| \le \kappa_G (r+1)^\kappa M^{\delta(G) r} \sum_{\substack{j \\ |j| \le 2N_1 - |k|}} \|(p^e)^j\,\hat{\Pi}(p^e)\|_{L^1} \tag{B.68a}$$

where N_1 is the total number of renormalization derivatives introduced, κ is the number of marginal C-forks, κ_G is a constant depending on G and $\delta(G) \le -1$.

b) If F is an R-fork and $r = -1$,

$$|G_{R,ren}| \le \kappa_G \sum_{\substack{j \\ |j| \le 2N_1 - |k|}} \sum_{i=0}^{n_0} \|\langle \partial_p e, \partial_p e\rangle^i (p^e)^j\,\hat{\Pi}(p^e)\|_{L^1} \tag{B.68b}$$

where $n_0 = 2|\mathcal{M}_F|$.

c) If F is a C-fork, then $G_{R,ren} = c \int \hat{\Pi}(p^e)\,dp^e$ where

$$|c| \le \kappa_G (r+1)^\kappa M^{-r \dim \Pi} , \tag{B.68c}$$

where $-2 \le \dim \Pi \le 0$.

<u>Remarks</u> 1. For expository purposes we have restricted ourselves to the case of a theory with a single scalar field, but the theorem extends easily, as in §2, to a relativistic theory involving any number of massive fields with dimensionless interaction.

2. We believe that with a more careful analysis it is possible to show that, as in Theorem 2.6, $\kappa_G \le c_0^{\ell(G)} \kappa!$.

3. In a scalar boson theory,

$$\dim \Pi = |k| + \dim \Phi = |k| + v_e \frac{d-2}{2} - d \qquad \text{(see (B.9))}$$

$$\geq |k| - 2 \ .$$

<u>Proof.</u> We rewrite (B.48) by integrating by parts with respect to α_ℓ, $\ell \in \mathcal{M}_0$, as described in (B.50) and (B.53). The resulting expression for $G_{R,ren}$ is a sum of terms of the form

$$c \tilde{\sum} \int e^{-im^2(\alpha)} U^{-d/2} \int dp^e \, e^{i\langle p^e, A^e p^e\rangle} \sum_j \beta_{\mathcal{M}_2} \partial_{\mathcal{M}_3} q_j \, \partial_{\mathcal{M}_4} \log U$$

$$\prod_i \partial_{\mathcal{M}_5^i} \log |A^e| \prod_i \langle \partial_p e, \partial_{\mathcal{M}_6^i} (A^e)^{-1} \partial_p e\rangle (p^e)^j \, \hat{\Pi}(p^e) \ . \qquad \text{(B.69)}$$

Here $\tilde{\sum}$ denotes the sum over scales and partition functions $\tilde{\chi}(\alpha,h)$, modified by the smoothing of χ_0 and χ_1 and by the integrations by parts; the sets $\mathcal{M}_2$, ..., $\mathcal{M}_6^i$ partition the different ways in which the derivatives $\partial^2_{\alpha_\ell}$, $\ell \in \mathcal{M}_1$, may act:

$$\mathcal{M}_1^2 = \mathcal{M}_2 \cup \mathcal{M}_3 \cup \mathcal{M}_4 \cup (\bigcup_i \mathcal{M}_5^i) \cup (\bigcup_i \mathcal{M}_6^i) \ ;$$

the factor $\beta_{\mathcal{M}_2}$ comes from ∂_{α_ℓ}'s acting on the partition functions $\tilde{\chi}_0(\alpha_\ell)$; the q_j's implicitly contain the effects of the integrations by parts with respect to α_ℓ, $\ell \in \mathcal{M}_{0,C}$.

We bound the factors in (B.69) in the obvious way: the oscillatory factors by 1, $U^{-d/2}$ by (B.30), $\partial_{\mathcal{M}_3} q_j$ by (B.59), $\partial_{\mathcal{M}_4} \log U$ and $\partial_{\mathcal{M}_5^i} \log |A^e|$ by (B.67) and $\partial_{\mathcal{M}_6^i} (A^e)^{-1}$ by (B.63). The result is

$$|\text{(B.69)}| \leq \kappa \tilde{\sum} \int \beta^{d/2}_{\mathcal{M}^c} \beta^3_{\mathcal{M}_{0,C}} \beta^2_{\mathcal{M}_1} \prod_f \tilde{\alpha}_f^{n_f/2} M^{-N_f^e(h_f - h_{\pi(f)})} \sum_j |\Pi|_{n_0,j}$$

where

$$|\Pi|_{n_0,j} = \sum_{i=0}^{n_0} \| \langle \partial_p e, \partial_p e\rangle^i (p^e)^j \, \hat{\Pi}(p^e) \|_{L^1} \ .$$

Now $\tilde{\alpha}_f^{n_f/2} = 1$ except at a 0-scale C-fork where $\tilde{\alpha}_f^{n_f/2} \le M^{-2} \max_{\ell \in \mathcal{M}_f} \alpha_\ell$. Hence

$$\beta_{\mathcal{M}_{0,C}} \prod_f \tilde{\alpha}_f^{n_f/2} \le 1$$

and so

$$|(B.69)| \le \kappa \tilde{\tilde{\sum}} \int \beta_{\mathcal{M}^c}^{d/2} \beta_{\mathcal{M}_0}^2 \prod_f M^{-N_f^e(h_f - h_{\pi(f)})} \sum_j |\pi|_{n_0,j} .$$

We can now integrate out α as in (B.38). For $d > 2$ the factors $\beta_{\mathcal{M}^c}^{d/2} \beta_{\mathcal{M}_0}^2$ give convergence of $\int d\alpha_\ell$ at $\alpha_\ell = \infty$ and we obtain

$$|(B.69)| \le \kappa \sum_h M^{(D^0(G_f) - N_f^e)(h_f - h_{\pi(f)})} |\pi|_{n_0,j} . \tag{B.70}$$

(B.69) simplifies considerably in cases a) and c) of the theorem.

<u>Case a):</u> Since $h_F > 0$, $\mathcal{M}_1 = \emptyset$ and so the bound (B.70) holds with $n_0 = 0$.

<u>Case c):</u> $\mathcal{M}_1 = \emptyset$; moreover, $t_F = 0$ so $x^e(t) = 0$ and the kernel of G is evaluated at $p^e = 0$. Thus the factor $e^{i\langle p^e, A^e p^e \rangle}$ and all terms with $j \ne 0$ are eliminated from (B.69).

<u>Case b):</u> When $h_F = 0$ we pull apart $R = 1 - L$ at F and estimate the 1-fork by (B.70) and the L-fork as in case c). When $h_F > 0$ the estimates of a) apply.

Finally, we estimate the sum over h in (B.70) as in Theorem 2.6 and thus establish (B.68abc).

We remark that our methods can be adapted to the case of massless particles to give a relativistic version of Theorem 6.5 for the IR-renormalized tree expansion. To the reader who has persevered to this point and whose energy is not exhausted, we leave the IR case as a homework exercise.

References

1. J. Schwinger (ed.), Quantum Electrodynamics, Dover, New York, 1958.

2. S. Weinberg, Phys. Rev. **118**, 838-849 (1960).
   Y. Hahn and W. Zimmermann, Commun. Math. Phys. **10**, 330 (1968).

3. N. Bogoliubov and O.S. Parasiuk, Acta Math. **97**, 227-266 (1957).

4. K. Hepp, Commun. Math. Phys. **2**, 301-326 (1966).
   Théorie de la Rénormalization, Springer Lecture Notes in Physics, Vol. 2, Springer, Berlin, 1969.
   Renormalization Theory, pp. 429-500 in Statistical Mechanics and Quantum Field Theory, ed. C. DeWitt and R. Stora, Gordon and Breach, New York, 1971.

5. W. Zimmermann, Commun. Math. Phys. **15**, 208-234 (1969).

6. G. Velo and A.S. Wightman (ed.), Renormalization Theory, D. Reidel, Dordrecht, Holland, 1976, in particular, the lectures of J.H. Lowenstein, BPHZ Renormalization, pp. 95-160.

7. K.G. Wilson, Phys. Rev. **D3**, 1818-1846 (1971) and **D6**, 419-426 (1972).
   J. Kogut and K.G. Wilson, Phys. Rep. **12** (1974) 75-200.

8. G. Benfatto, M. Cassandro, G. Gallavotti, F. Nicolò, O. Olivieri, E. Presutti, and E. Scacciatelli, Commun. Math. Phys. **59**, 143 (1978) and **71**, 95 (1980).

9. G. Gallavotti, Rev. Mod. Phys. **57**, 471-562 (1985).

10. G. Gallavotti and F. Nicolò, Commun. Math. Phys. **100**, 545-590 (1985) and **101**, 247-282 (1986).

11. P. Matthews and A. Salam, Rev. Mod. Phys. **23**, 311-314 (1951).

12. N. Bogoliubov and D. Shirkov, Introduction to the Theory of Quantized Fields, Interscience, New York, 1959.
    S. Schweber, An Introduction to Relativistic Quantum Field Theory, Harper and Row, New York, 1961.
    J.D. Bjorken and S.D. Drell, Relativistic Quantum Fields, McGraw-Hill, New York, 1965.
    J. Jauch and F. Rohrlich, The Theory of Photons and Electrons, Springer-Verlag, New York, 1976.

13. C. Itzykson and J.B. Zuber, Quantum Field Theory, McGraw-Hill, New York, 1980.

14. J. Magnen and R. Sénéor, Ann. Phys. **152**, 130-202 (1984).

15. K. Gawedzki and A. Kupiainen, Ann. Phys. **147**, 198-243 (1983), and Commun. Math. Phys. 92, 531-553 (1984).

16. C. de Calan and V. Rivasseau, Commun. Math. Phys. **82**, 69-100 (1981).

17. J. Polchinskii, Nuclear Phys. **B231**, 269-295 (1984).

18. K. Symanzik, Euclidean Quantum Field Theory, pp. 153-226 in Local Quantum Theory, ed. R. Jost, Academic Press, New York, 1969.
    J. Glimm and A. Jaffe, Quantum Physics, A Functional Integral Point of View, Springer-Verlag, Berlin, 1987.

19. G. Jona-Lasinio, Nuovo Cimento **L34**, 1790-1795 (1964).

20. A.S. Wightman, Orientation, pp. 1-24 in Ref. 6.

21. F.A. Berezin, The Method of Second Quantization, Academic Press, New York, 1966.

22. J.C. Ward, Phys. Rev. **78**, 1824 (1950).
Y. Takahashi, Nuovo Cimento **6**, 370 (1957).

23. W. Pauli and F. Villars, Rev. Mod. Phys. **21**, 434 (1949).

24. G. 't Hooft and M. Veltman, Nuc. Phys. **B44**, 189-213 (1972).
E. Speer, Dimensional and Analytic Renormalization, pp. 25-94 in Ref. 6.
P. Breitenlohner and D. Maison, Commun Math. Phys. **52**, 11-38 (1977).
L. Rosen and J. Wright, Dimensional Regularization and Renormalization of QED, preprint.

25. W.H. Furry, Phys. Rev. **51**, 125 (1937).

26. W. Zimmermann, The Power Counting Theorem for Feynman Integrals with Massless Propagators, pp. 171-184 in Ref. 6.

27. J. Feldman, J. Magnen, V. Rivasseau and R. Sénéor, Commun. Math. Phys. **98**, 273-288 (1985).

28. P. Blanchard and R. Sénéor, Ann. Inst. H. Poincaré **19**, 147-209 (1975).

29. C. de Calan, D. Petritis and V. Rivasseau, Commun. Math. Phys. **101**, 559-577 (1985).

30. G. Parisi, Phys. Rep. **49**, 215 (1979).
G. 't Hooft, Can we make sense out of Quantum Chromodynamics?, pp. 943-982 in The Why's of Sub-nuclear Physics, ed. A. Zichichi, Plenum Press, 1979.
B. Lautrup, Phys. Lett. **69B**, 109 (1977).

31. I. P. Goulden and D. M. Jackson, Combinatorial Enumeration, Wiley, New York (1983).

32. J. Feldman, J. Magnen, V. Rivasseau and R. Sénéor, Commun. Math. Phys. **100**, 23-55 (1985).

33. K. Osterwalder, Euclidean Green's Functions and Wightman Distributions, pp. 71-93, in Constructive Quantum Field Theory, ed. G. Velo and A. Wightman, Springer-Verlag, Berlin, 1973.

34. K. Osterwalder and R. Schrader, Commun. Math. Phys. **31**, 83-112 (1973) and **42**, 281-305 (1975).

35. J.-P. Eckmann, H. Epstein, and J. Fröhlich, Ann. Inst. H. Poincaré **25**, 1-35 (1976).

36. J. Lowenstein and E. Speer, Commun. Math. Phys. **47**, 43-51 (1976).

37. N. Nakanishi, Graph Theory and Feynman Integrals, Gordon and Breach, New York, 1970.